CARBONATES

SEDIMENTOLOGY, GEOGRAPHICAL DISTRIBUTION AND ECONOMIC IMPORTANCE

GEOLOGY AND MINERALOGY RESEARCH DEVELOPMENTS

Additional books in this series can be found on Nova's website
under the Series tab.

Additional e-books in this series can be found on Nova's website
under the e-book tab.

CHEMICAL ENGINEERING METHODS AND TECHNOLOGY

Additional books in this series can be found on Nova's website
under the Series tab.

Additional e-books in this series can be found on Nova's website
under the e-book tab.

GEOLOGY AND MINERALOGY RESEARCH DEVELOPMENTS

CARBONATES

SEDIMENTOLOGY, GEOGRAPHICAL DISTRIBUTION AND ECONOMIC IMPORTANCE

BAILEY A. HUGHES

AND

THOMPSON C. WAGNER

EDITORS

NOVINKA

New York

For permission to use material from this book please contact us:
Telephone 631-231-7269; Fax 631-231-8175
Web Site: http://www.novapublishers.com

NOTICE TO THE READER

The Publisher has taken reasonable care in the preparation of this book, but makes no expressed or implied warranty of any kind and assumes no responsibility for any errors or omissions. No liability is assumed for incidental or consequential damages in connection with or arising out of information contained in this book. The Publisher shall not be liable for any special, consequential, or exemplary damages resulting, in whole or in part, from the readers' use of, or reliance upon, this material. Any parts of this book based on government reports are so indicated and copyright is claimed for those parts to the extent applicable to compilations of such works.

Independent verification should be sought for any data, advice or recommendations contained in this book. In addition, no responsibility is assumed by the publisher for any injury and/or damage to persons or property arising from any methods, products, instructions, ideas or otherwise contained in this publication.

This publication is designed to provide accurate and authoritative information with regard to the subject matter covered herein. It is sold with the clear understanding that the Publisher is not engaged in rendering legal or any other professional services. If legal or any other expert assistance is required, the services of a competent person should be sought. FROM A DECLARATION OF PARTICIPANTS JOINTLY ADOPTED BY A COMMITTEE OF THE AMERICAN BAR ASSOCIATION AND A COMMITTEE OF PUBLISHERS.

Additional color graphics may be available in the e-book version of this book.

Library of Congress Cataloging-in-Publication Data
Carbonates : sedimentology, geographical distribution and economic importance / editors, Bailey A. Hughes and Thompson C. Wagner.
      pages cm
  Includes index.
  ISBN: 978-1-62948-178-4 (soft cover)
  1. Carbonates. 2. Marine sediments. I. Hughes, Bailey A., editor of compilation. II. Wagner, Thompson C., editor of compilation.
  QE471.15.C3C393 2011
  549'.78--dc23
                        2013036362

Published by Nova Science Publishers, Inc. † *New York*

CONTENTS

PREFACE

In this book, the authors present current research in the study of the sedimentology, geographical distribution and economic importance of carbonates. Topics discussed in this compilation include the alternative procedures for the synthesis of linear carbonates from alcohols and carbon dioxide; functional polymers based on carbonates obtained from CO2; an experiment using soil micromorphology and image analysis for physical redistribution of calcium carbonate in soil pore systems; and the types of petroleum reservoirs in carbonate sediments of the Russian Basin.

Chapter 1 - The sequestration of CO_2 into organic carbonates had attracted considerable attention from the perspective of green chemistry. Among them, the aliphatic alkyl carbonates, Dimethyl Carbonate (DMC) and Diethyl Carbonate (DEC), are interesting products due to its nontoxicity and excellent solubility, besides of being readily biodegradable into CO_2 and alcohols. Additionally, DMC and DEC are versatile intermediates for organic synthesis, and can be used as octane enhancers for gasoline and diesel fuels, decreasing soot and CO emissions. Several synthetic routes for the synthesis of dialkyl carbonates are available; however, these methodologies involve the use of highly toxic compounds such as phosgene, carbon monoxide, hydrochloric acid, and nitric oxide. An alternative route for the production of linear carbonates is the reaction of carbon dioxide with a linear alcohol, methanol or ethanol for producing DMC or DEC, respectively. In this chapter the authors first summarize the state-of-the-art on the direct synthesis of DMC and DEC from CO_2 and Methanol/Ethanol. Then, the authors present their results in the synthesis, characterization and evaluation of Cu-Ni catalysts supported on activated carbon for the synthesis of DMC and DEC at moderate conditions (i.e., P < 15 bar, T < 150°C). A detailed discussion on the role of the Ni-Cu alloy formed on the catalytic activity is also included.

Chapter 2 - This chapter describes synthesis and application of various functional polymers from carbonates obtained by the reaction of epoxides and CO_2. Main subjects are polyhydroxyurethanes prepared via polyaddition of cyclic carbonates and amines, and polymers bearing cyclic carbonate moieties prepared from glycidyl methacrylate (GMA) and CO_2.Polyhydroxyurethanes, polyurethanes bearing hydroxyl groups in the side chains, are prepared via polyaddition of bifunctional carbonates with diamines. This polyaddition is chemo-selective, and various functional monomers and polymerization conditions are available even though the primary reaction is a nucleophilic addition. Transformation of hydroxyl side chains also introduces various functional groups. The potential applications involve paints, packaging materials, biomaterials, and electronic devices. Polymers bearing carbonate moieties in the side chain are potentially applicable as optical materials, polymeric electrolytes, adhesives, and paints. One of the most facile methods is the reaction with GMA and CO_2, in which three pass ways are possible.The synthetic methods are polymerization of a methacrylate bearing carbonate moieties obtained by the reaction of GMA and CO_2, reaction of polyGMA and CO_2, and concurrent polymerization and CO_2 fixation via radical polymerization of GMA under CO_2 atmosphere.

Chapter 3 - Calcium carbonate sedimentation plays an important role in the soil structure development. Soil structure is a key factor for the soil functions, such as sustaining plant productivity, regulating and partitioning water and solute flow, and maintaining civil engineering works. In order to improve soil structure the calcium carbonate is added to the soil by means liming procedure in order to improve soil chemical and physical properties, neutralizing soil acidity and increasing activity of soil bacteria, and improving the soil structure. A large bulk of scientific literature addresses the relations between the forms of carbonate redistribution in the development of calcic and petrocalcic horizons and the water regime of calcareous hydromorphic soils, but little is still known concerning the underlying physical mechanisms of the effect of carbonate sedimentation on soil pore system. In this work the authors attempted to investigate physical mechanisms of soil pore development as consequence of the addition of calcium carbonate ($CaCO_3$) on two soils with different shrinkage-swelling capacity subjected to several wetting and drying cycles. Analysis was conducted using 2D soil image analysis and soil micromorphology. The author's results showed changes in the pore size distribution, in some cases very large, and allowed the identification of specific mechanisms of pore modification induced by micrite pedofeatures produced by the mobilization in suspension of $CaCO_3$. These physical mechanisms were triggered by $CaCO_3$ segregations, which induced a pore size redistribution fragmenting the pore space, and $CaCO_3$ coatings, which seemed to induce a cumulative effect on porosity cementing the walls of newly-formed pores in the soil samples with high shrinkage-swelling capacity.

Our results, even if obtained on experimental samples, give a contribution in the understanding of the physical role of $CaCO_3$ pedofeatures in pore formation in soils in field and show the need to reassess physical simulation tests in order to quantitatively investigate combined effects of factors influencing soil structure formation.

Chapter 4 - In petroleum basins of the Russian Federation oil and gas fields in carbonate reservoirs have been discovered in rocks ranging from the Riphean to the Eocene. giant fields, are controlled by reefs. Depending on the paleoclimatic zone, the seals are composed of salt or, rarely, of shale. The largest amount of the fields are found in cratonic carbonate formations deposited under arid climatic conditions. Regional seals are formed by salt, anhydrite, and dolomicrite. Multilayer reservoirs predominate, but massive reservoirs are also common. The distribution of reservoir types and their quality are strongly uneven. A large number of fields, including giant fields, are controlled by reefs. Massive reservoirs predominate, but the distribution of porosity and localization of zones of improved reservoir properties are variable and controlled by the morphogenetic types of the reefs- Carbonate formations deposited under humid climatic conditions contain much less hydrocarbon reserves. The seals are generally composed of shale. The reservoirs are stratal, rarely multilayer. The fields are commonly small. A number of fields, some of them highly productive, are present in Upper Cretaceous carbonate rocks of the North Caucasus region. The carbonates consist of remain of planktonic organisms. Seals for the hydrocarbon pools are composed of shale. The reservoirs are massive and layered-massive. Fractures and stylolites play a leading role in controlling the reservoir properties.

In: Carbonates   ISBN: 978-1-62948-178-4
Editors: B.A. Hughes, T.C. Wagner   © 2013 Nova Science Publishers, Inc.

Chapter 1

ALTERNATIVE PROCEDURES FOR THE SYNTHESIS OF LINEAR CARBONATES FROM ALCOHOLS AND CARBON DIOXIDE

Felipe Bustamante, Oscar Arbeláez†, Andrés Orrego‡ and Aída Luz Villa§*

Environmental Catalysis Research Group, Chemical Engineering
Department, Universidad de Antioquia, Medellín, Colombia

ABSTRACT

The sequestration of CO_2 into organic carbonates had attracted considerable attention from the perspective of green chemistry. Among them, the aliphatic alkyl carbonates, Dimethyl Carbonate (DMC) and Diethyl Carbonate (DEC), are interesting products due to its nontoxicity and excellent solubility, besides of being readily biodegradable into CO_2 and alcohols. Additionally, DMC and DEC are versatile intermediates for organic synthesis, and can be used as octane enhancers for gasoline and diesel fuels, decreasing soot and CO emissions. Several synthetic routes for the synthesis of dialkyl carbonates are available; however, these methodologies involve the use of highly toxic compounds such as phosgene, carbon monoxide, hydrochloric acid, and nitric oxide. An

---

[*] Corresponding authors. fbustama@udea.edu.co. Tel.: (+57)-42198535; Fax: (+57)-42119028.
[†] oscarfelipe3@yahoo.es; Tel.: (+57)-42198535.
[‡] che.andresorrego@gmail.com; Tel.: (+57)-42198535.
[§] alvilla@udea.edu.co; Tel.: (+57)-42198535.

alternative route for the production of linear carbonates is the reaction of carbon dioxide with a linear alcohol, methanol or ethanol for producing DMC or DEC, respectively. In this chapter we first summarize the state-of-the-art on the direct synthesis of DMC and DEC from CO_2 and Methanol/Ethanol. Then, we present our results in the synthesis, characterization and evaluation of Cu-Ni catalysts supported on activated carbon for the synthesis of DMC and DEC at moderate conditions (i.e., P < 15 bar, T < 150°C). A detailed discussion on the role of the Ni-Cu alloy formed on the catalytic activity is also included.

1. INTRODUCTION

Carbon dioxide, a greenhouse gas, is considered one of the main contributors to global warming. Nowadays, a number of technologies for recovering, utilization and sequestering carbon dioxide are being developed with a view to increasing their efficiency, reliability and cost [1]. One of these methods is the geological sequestration of CO_2 by the carbonation with minerals (calcium, magnesium, sodium and potassium) to form stable compounds such as inorganic carbonates; however, these methods are excessively slow [2]. The sequestration of CO_2 into organic carbonates, on the other hand, has attracted considerable attention from the perspective of green chemistry. Among the organic carbonates, the aliphatic alkyl carbonates, Dimethyl Carbonate (DMC) and Diethyl Carbonate (DEC) (figure 1), are very interesting products due to their nontoxicity, excellent solubility, and readily biodegradability into CO_2 and alcohols [3]. Additionally, they are versatile intermediates for organic synthesis used as methylating and carbonylating agents [4] and can be used as octane enhancers for gasoline and diesel fuels, decreasing the emissions of particulate matter and CO [5].

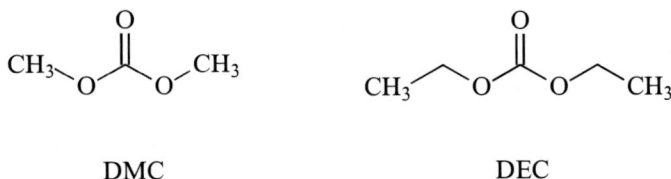

DMC   DEC

Figure 1. DMC and DEC structures.

Several synthetic routes towards the dialkyl carbonates, including alcoholysis (methanolysis [7] or ethanolysis [8]) of phosgene, oxidative carbonilation of methanol [9] or ethanol [10], and methyl [11] or ethyl nitrite carbonylation [12], have been used. However, these methodologies have safety concerns due to the use of extremely toxic compounds, such as phosgene. An altenative route for the production of linear carbonates is the reaction of carbon dioxide with a linear alcohol, (figure 2) methanol [13] or ethanol [14], to produce DMC or DEC, respectively.

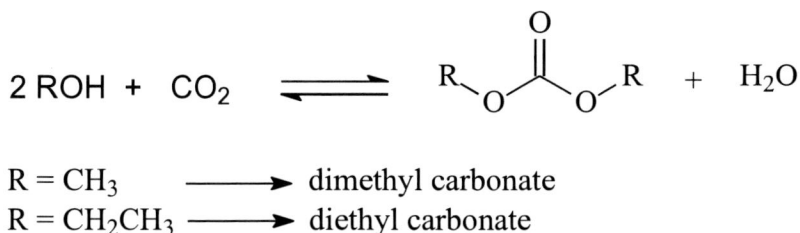

$$2\ ROH\ +\ CO_2 \rightleftharpoons \underset{O\quad\quad O}{R} \overset{\displaystyle O}{\underset{}{\big\|}} R\ +\ H_2O$$

R = CH$_3$ $\longrightarrow$ dimethyl carbonate

R = CH$_2$CH$_3$ $\longrightarrow$ diethyl carbonate

Figure 2. Synthesis of linear carbonates from carbon dioxide and alcohols.

Due to its very low equilibrium conversion, this reaction is customarily conducted at low temperature and very high pressures. Therefore, efficient catalysts as well as moderate conditions are required for the production of carbonates from alcohols and CO_2. In fact, a gas-phase process carried out at moderate conditions would facilitate process control, catalyst recovery, and would reduce both operational and capital costs. Several catalytic systems have already been reported for the direct synthesis of DMC from carbon dioxide and methanol. Most of these catalysts have been tested at very high pressures (above 4 MPa). In regards to the direct synthesis of DEC, fewer studies are available in the open literature.

In this chapter we first summarize the state-of-the-art on the direct synthesis of DMC and DEC from CO_2 and Methanol/Ethanol. Then, we present our results in the synthesis, characterization and evaluation of Cu-Ni catalysts supported on activated carbon for the synthesis of DMC and DEC at moderate conditions (i.e., P < 15 bar, T < 150°C). A detailed discussion on the role of the Ni-Cu alloy formed on the catalytic activity is also included.

2. DIRECT SYNTHESIS OF LINEAR CARBONATES FROM ALCOHOLS AND CO_2

2.1. Catalytic Systems for DMC Production

The direct synthesis of DMC by reaction between CO_2 and methanol has been studied since the late 1970's as an attractive route to replace traditional synthesis methods, such as phosgenation or oxidative carbonylation of methanol. Due to its unfavorable chemical equilibrium, which is improved only at low temperature and/or high pressure, most studies have been conducted in the liquid phase and/or under high pressure conditions (supercritical conditions) showing disadvantages related with to the high cost of the starting materials, difficulty in the catalyst-product separation due to the homogeneous nature of the catalyst, hydrolysis of the carbonate, and the rapid catalyst deactivation or decomposition. Recently, efforts have been made in developing heterogenous catalysts.

A number of reviews of catalytic systems for the direct synthesis of DMC from carbon dioxide and methanol are available in the literature [15-20]. Some relevant findings are summarized below.

In early works, particular attention was given to the use of homogeneous catalysts based in Sn (IV) and Ti (IV) tetra-alkoxides. The first reports on organometallic complexes catalysts for this reaction were published by a Sakai *et al.* [21] and Yamasaki *et al.* [22, 23] in the late 1970s, who obtained conversions above those expected from the reaction stoichiometry. Later, Kizlink *et al.*, between 1993 and 1995 [24–26], used dialkyltin dialkoxides, $BuSn(OMe)_2$, thus improving the catalytic activity of this reaction at the expense of working at high pressure (4-8 MPa). After Kizlink, organometallic compounds based on Sn(IV) such as $n-R_2Sn(OMe)_2$ (R = Me, Bu) [27], series of n-butyl(alkoxy) stannanes [28–31], dibutyltin oxide [32], decakis(diorganotin(IV)) oxoclusters [33] and immobilized [34–36], were applied successfully; DMC yield reported was between 30 and 88% in these systems. The high performance has been partly attributed to the fact that the direct synthesis becomes thermodynamically favorable as the system pressure increases.

Some strategies, such as the *in situ* remotion of water from the reaction medium or the inclusion of dehydrating agents, have been explored for overcoming the low equilibrium conversions in the homogeneous systems. For example, the reaction has been carried out in the presence of dehydrating agents, additives, or promoters such as ortho esters (e.g., orthoacetate, $Si(OMe)_4$) [37, 38], acetals [19, 39], 2,2-dimethoxypropane [40], or molecular sieves [41]) to minimize the water concentration in the reaction system. However, the method requires expensive orthoesters as starting materials. In addition, it is difficult to regenerate the orthoesters from esters and alcohols. Employing other dehydrating agents, such as dicyclohexyl carbodiimide (DCC), $Si(OMe)_4$ and Mitsunobu's reagent, poses the same problem. Hence, more easily available and recyclable organic dehydrating agents must be developed. $Ni(CH_3COO)_2$ [42−44], carbodiimides [45], tantalum alkoxides [46] and niobium complexes [47−49] also catalyze the reaction in homogeneous phase.

On the other hand, basic heterogeneous catalysts (K_2CO_3, CH_3OK, KOH) [38, 50−53], solid acid−base catalysts such as zirconia [54, 55], ceria-modified zirconia [56−60], acidic compounds $H_3PW_{12}O_4$−ZrO_2 [61], phosphoric acid H_3PO_4−ZrO_2 [62, 63] or ZrO_2−KCl [64], promote catalytic activity. Heteropolyacids ($H_3PW_{12}O_{40}$−$Ce_xTi_{1-x}O_2$ and $H_3PW_{12}O_{40}$−$Ce_xZr_{1-x}O_2$) [65−67], CeO_2 [68], Al_2O_3−CeO_2 [69], $Ga_2O_3/Ce_xZr_{1-x}O_2$ [70] metal oxide−$Ce_{0.6}Zr_{0.4}O_2$ [71], ZrO_2−MgO [72], ZrO_2/SiO_2, and SnO_2/SiO_2 [73] are also active materials for the DMC formation from methanol and CO_2. In this group of catalysts, the reaction was carried out at pressures above 4 MPa, and, similarly to those using homogeneous catalysts, batch operations at high pressure can increase operational cost in the process. Therefore, a gas-phase process carried out at moderate conditions would greatly facilitate process control, catalyst recovery, and would reduce both operational and capital costs.

Few catalytic studies have been reported in batch operation at low pressure (0.12−0.5 MPa), where heterogeneous catalysts such as $Cu_{1.5}PMo_{12}O_{40}$ [74], CeO_2 with acetonitrile as dehydrating agent [75], and heteropolyoxometalates [76, 77] have been tested without showing a significant increase of DMC yield compared to the gas-phase reaction in continuous reaction system.

Heterogeneous catalysts reported for the gas-phase synthesis at low pressure (0.1−1.2 MPa), moderate temperature (353−500 K) and continuous flow, include Ni−Cu/MoSiO(VSiO) [78], Cu−KF/MgSiO [79], H_3PO_4−V_2O_5 [80] Cu−Ni/VSO [81], Cu−(Ni,V,O)/SiO_2 and with photo-assistance [82], Cu−Ni/C [83], (C = graphite [84], MWCNTs [85], activated carbon [86], TEG [87], graphite oxide [88], GNS [89]), Rh/ZSM-5 [90], $CuCl_2$/AC [91], $Co_{1.5}PW_{12}O_{40}$ [92], Cu−Ni/diatomite [93] and Cu-Ni/molecular sieve [94]. Yield in these systems appear to be lesser than in the homogeneous reaction.

Table 1. Heterogeneous catalysts reported for gas-phase synthesis of DMC

Entry	Catalyst	Conditions	DMC yield based on MetOH	Author (year)
1	Ni−Cu/MoSiO(VSiO)	140°C, 1 bar	7.08%	Zhong et. al. (2000)
2	Cu-KF/MgSiO	130°C, 10 bar	2.95%	Li and Zhong (2003)
3	$H_3PO_4-V_2O_5$	140°C, 6 bar	1.80%	Wu et. al. (2005)
4	Cu-Ni/VSO	140°C, 6 bar	1.10%	Wu et. al. (2006)
5	Cu−(Ni,V,O)/SiO$_2$ with photoassistance	120°C, UV irradation	4%	Wang et. al. (2007)
6	Cu-Ni/AC	110°C, 12 bar	5.31%	Bian et. al. (2009)
7	Rh/ZSM-5	120°C, 1 bar	24.9 % (X_{MeOH} = 41%)	Almusaiteer (2009)
8	$Co_{1.5}PW_{12}O_{40}$	200°C, 1 bar	6.57%	Aouissi et. al. (2010)
9	CuCl$_2$/AC	120°C, 12 bar	DMC rate = 4.77 mmol h^{-1}	Bian et. al. (2010)
10	Cu-Ni/Novel Graphene Nanosheet	120°C, 12 bar	(X_{MeOH} = 9.22%)	Bian et. al (2011)
11	Cu−Ni/diatomite	120°C, 12 bar	5.93%	Chen et. al. (2012)
12	Cu−Ni/molecular sieve	120°C, 11 bar	5.0%	Chen et. al. (2012)

According to Table 1, DMC yields in the direct synthesis in the gas phase are below of 7%, with exception of the work by Almusaiteer *et. al.* [19], entry 7. However, a DMC yield of 24.9% and 41% methanol conversion may surpass the equilibrium conversion of the gas-phase reaction at those conditions (see [95]). From Table 1 it is also evident that Cu-Ni over activated carbon is an effective catalyst in the formation of DMC. In this catalyst, metal phase Cu, Ni and Cu-Ni alloy plays an important role in the activation of reactants. Furthermore, the properties of activated carbon such as low cost, high chemical and physical stability, and outstanding electronic transport properties, make this material prominent catalyst both for DMC and DEC synthesis.

2.2. Catalytic Systems for DEC Production

Few studies on the direct synthesis of diethyl carbonate from carbon dioxide and ethanol have been reported. Fujita *et al.* [96] conducted the reaction using methanol and ethanol at supercritical conditions in the presence of an heterogeneous catalysts (CH_3CH_2I-K_2CO_3); ethanol was found to have less reactivity than methanol for the carbonate synthesis; additionally, the activity of the catalyst was very low, probably because the ethyl iodide was behaving more like as a reactant than as a catalyst for the reaction. Based on these results, Gasc *et al.* [97] synthesized diethyl carbonate in both one-pot (direct method) and two-step (indirect method) reactions using K_2CO_3 as catalyst, and ethanol, ethyl iodide and CO_2 as reactants; the authors reported that the one-pot method gave higher DEC selectivity (diethyl ether is the main by-product) and DEC yield than the two-step. Yoshida *et al.* [68], evaluated CeO_2 as catalyst in the synthesis of dimethyl carbonate, diethyl carbonate and ethyl methyl carbonate (EMC) from the corresponding alcohols and CO_2; the rate of DEC production was much lower than that of DMC and EMC production. Wang *et al.* [98] studied the performance of a cerium-modified zirconia, $Ce_xZr_{1-x}O_2$, as catalyst in the synthesis of DEC, showing that the presence of strong acid and basic sites in the catalyst are unfavorable for the formation of the carbonate; additionally, the authors pointed out that the experimental conditions can increase operational cost in the process and the reaction conditions are more difficult to control.

The results of the synthesis of DEC are summarized in Table 2. All of the catalytic systems have been evaluated in the liquid-phase synthesis and at high pressure; additionally, these catalysts are not attractive in terms of green and sustainable chemistry. Consequently, the development of new heterogeneous catalysts for the gas-phase synthesis, with the concomitant advantages of low cost and easier catalyst recovery and process control, is required.

Table 2. Heterogeneous catalysts reported for DEC synthesis

Entry	Catalyst	Conditions	DEC yield (ethanol basis)	Author (year)
1	CH_3CH_2I-K_2CO_3	70°C, 80 bar	not reported	Fujita *et al.* (2001)
2	K_2CO_3	110°C, 80 bar	5 %	Gasc *et al.* (2009)
3	CeO_2	170 °C, 50 bar	0.187 %	Tomishige *et al.* (2006)
4	$Ce_xZr_{1-x}O_2$	140°C, 80 bar	0.1 %	Wang *et al.* (2009)

2.3. Cu-Ni bimetallic Catalyst

From the catalytic systems mentioned before, the Cu-Ni bimetallic catalyst supported on different carbon materials, which was patented in 2008 [99], appears to be the most promising for obtaining dimethyl carbonate in the gas phase at moderate conditions of pressure and temperature [78- 94].

Due to their vicinity in the periodic table copper and nickel have very similar atomic configuration, atomic weight, electro-negativity, atomic radius, and valence. Besides, both metals have the same crystal structure (face centered cubic, *fcc*), their lattice constants differing only by 2.5% [100]. Moreover, the two elements are completely miscible in both liquid and solid phase [101].

Cu and Ni-based catalysts are generally known to be very active in different kind of reactions where CO_2 and methanol/ethanol are involved as reactants. Processes such as steam reforming of methanol (SRM), partial oxidation of methanol (POM) and oxidative steam reforming of methanol (OSRM), are some examples where both metal surfaces can activate methanol. On the other hand, studies on CO_2 adsorption on Ni-based surfaces during the dry reforming of methane or the synthesis of methanol from CO_2 hydrogenation have been reported in the literature [102-106]. Furthermore, exciting results have been reported when Cu and Ni coexist in the same catalyst. For instance, copper is one of the most widely used catalysts in the synthesis of methanol from CO_2 and hydrogen, but recent studies have shown that Ni-doped Cu surface or Cu-Ni alloys can be up to 40 times more reactive [107-110].

In the experimental part of this chapter, a suitable method for the preparation of monometallic and bimetallic samples supported on activated carbon (AC) and for the evaluation of the effect of the contribution of either metal, i.e., copper and nickel, on the direct synthesis of DMC/DEC from methanol/ethanol and CO_2, is shown. In fact, up to date the role of each metal and the chemical nature of the active sites for this reaction have not been studied in detail. Therefore, several characterization techniques, including BET (N_2 adsorption-desorption), Chemical analysis, XRD (X-Ray Diffraction), TPR (Temperature Programmed Reduction), CO pulse chemisorption, SEM (Scanning Electron Microscopy) and TGA (Thermogravimetric Analysis), were used in order to get insight into the catalyst structure. Moreover, the characterization of the most active catalysts, Cu-Ni (2:1)/AC and Cu-Ni (3:1)/AC was compared to Cu and Ni monometallic samples to establish the differences of the bimetallic material and the individual metals.

The characterization and reaction tests presented in this chapter aim at showing that a synergistic effect between Cu and Ni play an important role in catalytic activity and, in the other hand, at evidencing the link between the synergistic effect with the possible formation of Cu-Ni alloy in the catalyst.

The experimental conditions (13 bar and 90 °C) were selected close to the dew point of the quaternary gas mixture according to the studied presented in [95], in order that both products (DMC/DEC) and reactantas (methanol/ ethanol) be close to the equilibrium conditions in the actual reaction system without partial condensation of the reaction mixture.

3. EXPERIMENTAL CONDITIONS

3.1. Catalyst Preparation

The activated carbon (AC) used as support was pretreated following the method suggested by Bian et al. 2009 [83]. Basically, AC was mixed with a 2 M HCl solution for 12 h under reflux; then, the solid was filtered, washed with distilled water and dried in an oven at 110°C for 12 h; afterwards, the solid was stirred with a 4 M H_2SO_4 solution for 4 h, filtered, washed with distilled water, dried at 100 °C, and stored in a dessicator before its use as a support. Cu-Ni/AC catalysts with a nominal metal loading of CuO + NiO = 20 wt% and different Cu/Ni molar ratios were prepared by the wetness impregnation method. The metal precursors, $Cu(NO_3)_2\cdot3H_2O$ (Carlo Erba, 99,5%) and $Ni(NO_3)_2\cdot6H_2O$ (Carlo Erba, 99,5%), were dissolved in aqueous ammonia (J.T. Baker 29%) by stirring; then, the cooper and nickel solution was added to the pretreated AC. The resulting mixture was stirred at ambient temperature for 24 hours, followed by rotoevaporation (100 °C, 35 rpm, 350 mm Hg) and drying at 90°C in an oven. Finally, the samples were calcined at 600 °C for 3 hours in N_2 flow with a heating rate of 0.5°C/min, and then activated at 600°C for 3 hours in 5% H_2/Ar flow with a heating rate of 0.5°C/min.

3.2. Catalyst Characterization

The surface area of prepared catalysts was measured by N_2 physisorption, using the single point method. Prior to analysis, the sample was outgassed at

250 °C for 0.5 h in a stream of N_2/He with a heating rate of 0.5°C/min. Chemical composition (Cu and Ni wt %) was determined by atomic absorption in a Thermo Electron spectrometer (model S4); the sample (0.05 g) was dissolved in a 3:1 hydrochloric acid/nitric acid solution. The XRD diffraction analysis was carried out on a Bruker Focus with $5 \leq 2\theta \leq 70°$, using CuKα1 radiation operated at 45 kV, 40 mA and 0.014 step size, using a counting time of 1 s per point. The obtained diffractograms were compared to those of known compounds from JCPDS (Joint Committee of Powder Diffraction Standards) index. H_2-TPR experiments of calcined catalyst samples were performed in a Micromeritics AutoChem II 2920 apparatus. 50 mg samples were pretreated at 5°C /min to 250 °C for 1 h in flowing helium (70 mL/min), and then cooled to 40 °C. Thereafter, samples were heated to 800 °C using 5% H_2/Ar (70 mL/min) at 8 °C /min. The signals of H_2 consumption were continuously monitored by a thermal conductivity detector (TCD). The SEM images were taken with a JEOL JSM-6490 using an accelerating voltage of 20 kV.

3.3. Catalyst Evaluation

The catalytic activity of synthesized Cu-Ni/AC catalysts was evaluated in a continuous stainless steel tubular fixed-bed reactor with an inner diameter of 7 mm. The set-up is shown in Figure 3: a stream of CO_2/helium was directed to a stainless steel bubbler containing the liquid alcohol at 90°C and 13 bar (4); the resulting mixture (5) was admitted to the reactor (8), which was previously loaded with 0.5 g of catalyst; additionally, inlet concentrations were measured by using the reactor bypass (6).

Reactants and products were analyzed online using a MS, QMS Thermostar 200 Pfeiffer. DEC, DMC, diethyl and dimethyl ether concentrations were evaluated by monitoring mass signals m/e⁻ = 91, 59, 74, and 47, respectively. The activity of the catalyst (in terms of TOF) was evaluated with Equation 1

$$TOF = \frac{mol\,carbonate}{mol\,active\,site\,x\,h} \tag{1}$$

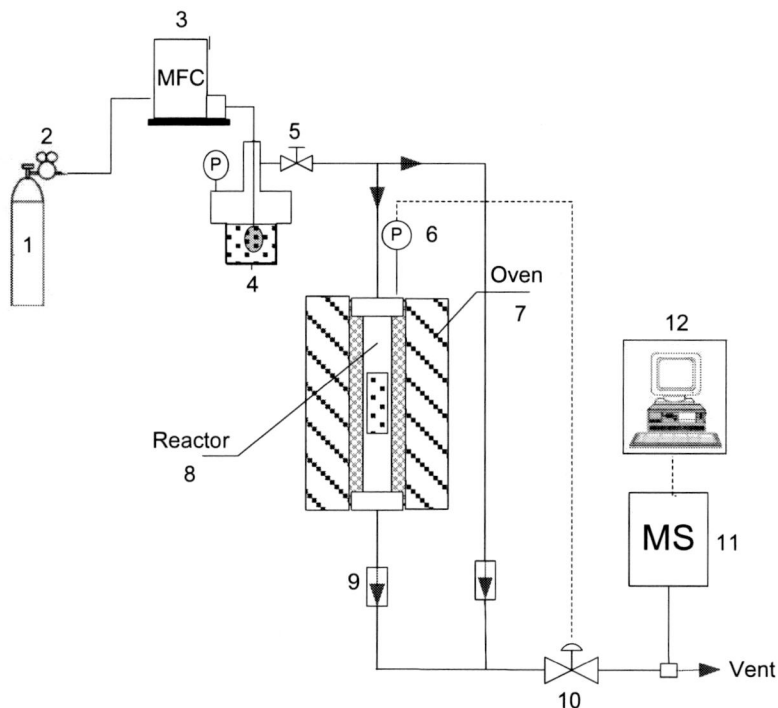

Figure 3. Reaction setup for gas phase synthesis of DMC/DEC from methanol and CO_2. (1) CO_2/He gas mixture, (2) pressure reducer, (3) mass flowrate controller (MFC), (4) bubbler containing methanol, (5) valve, (6) pressure gauge, (7) Oven (Temperature Control), (8) tubular fixed-bed reactor (7 mm I.D), (9) check valve, (10) Control Valve, (11) MS Spectrometer (ThermoStar QMS 200, Pfeiffer), (12) Data Processor.

4. RESULTS

4.1. Catalytic Activity Results

Figure 4 shows the catalytic activity, expressed as TOF (h^{-1}), of the monometallic and bimetallic catalysts at 90 °C and 13 bar; it was confirmed that the support was not active for the reaction (results not shown). The activity of the monometallic Ni catalysts is very low for both carbonates, whereas Cu displays high activity for DEC and low activity for DMC. Moreover, the monometallic catalysts are significantly more active in the synthesis of DEC than in DMC synthesis. Furthermore, the strong synergestic effect observed with all the bimetallic catalysts in DMC synthesis is not completely replicated with DEC; in fact, activity of the bimetallic catalysts with the higher nickel loadings, i.e., 1-1, 1-2 and 1-3, is lower than that of the monometallic copper catalyst. The highest TOF for DMC and DEC was obtained with Cu-Ni/AC ratio of 2:1 and 3:1, respectively. This increased catalytic activity is probably related to the formation of a Cu-Ni alloy formation observed by XRD (see below).

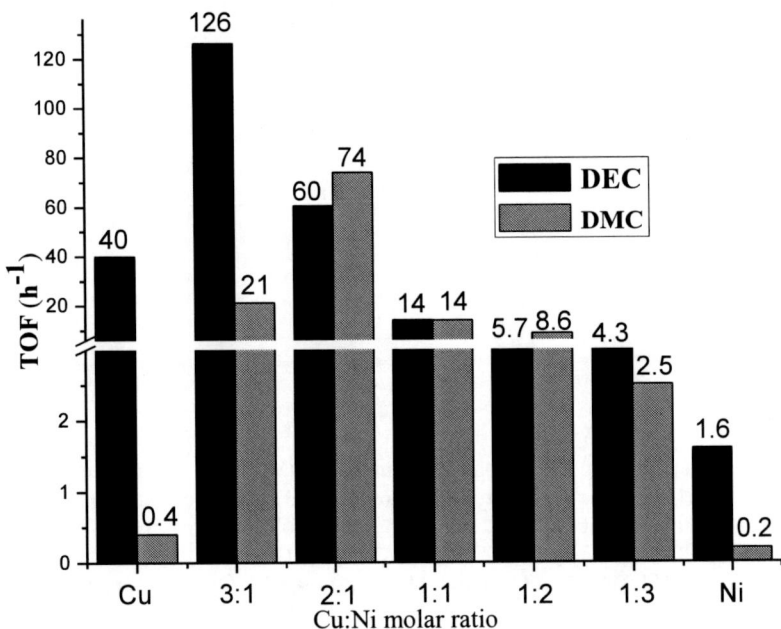

Figure 4. TOF for Cu-Ni/AC catalysts tested in the direct synthesis of DMC and DEC. Reaction conditions: 90°C, 13 bars, 0.5 g catalysts

4.2. Catalyst Characterization

4.2.1. Surface Area, Metal Dispersion, and Metal Loading Results

The surface area measurements, metal dispersion, and metal loading of the catalysts tested are given in Table 3.

Table 3. Surface area, metal dispersion and metal loading of the catalysts tested

Catalyst	Cu	CuNi 3:1	Cu-Ni 2:1	Cu-Ni 1:1	Cu-Ni 1:2	Cu-Ni 1:3	Ni
Surface area BET (m^2/g)	523	670	692	656	617	686	627
Metal dispersion (%)	0,1	0,02	0,04	0,13	0,25	0,55	0,1
wt % Cu	12,90	11,03	8,66	7,17	4,45	3,59	-
wt % Ni	-	2,70	2,79	4,68	5,57	7,68	9,43

Surface area of the catalysts is 10-19% lower than that of the activated carbon support (764 m^2 / g), which could be associated to the high temperature of calcination and reduction (500 ° C and 600 ° C, respectively) of the samples. However, a definite trend of surface area with Cu-Ni molar ratio is not observed. The lowest metal dispersion, on the other hand, observed for the catalyst with 2:1 and 3:1 Cu:Ni ratios, suggests that the presence of Cu and Ni as individual metal species is very low in these samples (Cu and Ni metallic species are the sites for carbon monoxide chemisorption). Oppositely, Cu and Ni would be present mostly as monometallic species in the 1:2, 1:2 and 1:3 bimetallic samples, as suggested by the larger metal dispersion measured using CO as probe molecule for chemisorption. Therefore, the low metal dispersion would be related to the synergestic effect observed in some bimetallic samples.

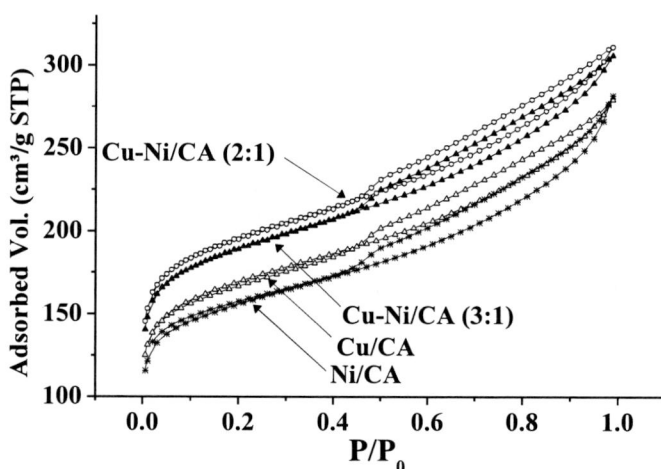

Figure 5. Adsorption-desorption isoterms of N$_2$ on Cu-Ni/CA samples.

Figure 5 shows the N_2 adsorption–desorption isotherms of the most active bimetallic catalysts, along with the monometallic catalysts. Bimetallic catalysts (Cu-Ni 3-1 and 2-1 molar ratios) exhibited an improved adsorption capacity, which is consistent with their larger surface areas compared to the monometallic catalysts (see Table 3). All isotherms are of type IV with hysteresis loops (IUPAC classification H4), which are usually observed in mesoporous solids with slit-shaped pores of non-uniform sizes and shapes, and important contribution of micropores. Adittionally, no significant changes in the morphology of activated carbon were observed in any sample (not shown).

4.2.2. X-ray Diffraction Results

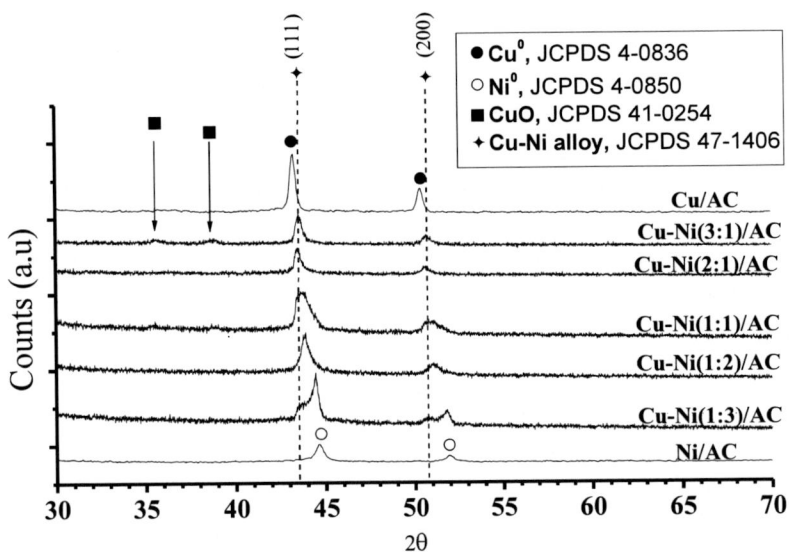

Figure 6. XRD of mono and bimetallic catalyst samples.

Figure 6 shows the X-ray diffraction patterns of mono and bimetallic copper-nickel catalysts. Monometallic samples Cu/AC and Ni/AC show peaks corresponding to Cu^0 ($2\theta = 43.3°$, $50.4°$, JCPDS file No. 4-0836) and Ni^0 ($2\theta = 44.5°$, $51.8°$, JCPDS file No. 4-0850) indicating the formation of crystalline metal particles. Moreover, a shift in the main diffraction peaks corresponding to the individual metals as Cu:Ni molar ratio increases, is evidenced. Moreover, XDR also indicate that Cu(111) and Ni(111) peaks of Cu^0 and Ni^0 metals decrease as the diffraction peak at $2\theta = 43.7°$, corresponding to Cu-Ni alloy, increases, to the point where a unique peak is seen (for samples with

Cu:Ni ratio of 2:1 and 3:1), suggesting that Ni and Cu atoms were well mixed at atomic level [114]. In fact, samples of Cu:Ni molar ratio of 2:1 and 3:1 show well-defined diffraction peaks that can be associated to a cubic phase Cu-Ni alloy (2θ = 43.7°, 50.9°, JCPDS file No-47-1406) [111-113]; oppositely, the presence of the Cu-Ni alloy in other bimetallic samples was not completely identified. The formation of cubic phase Cu-Ni alloy may be attributed to the pyrolysis and reduction steps, where the catalyst is activated prior to reaction. The presence of divalent copper oxide, i.e., CuO, (2θ = 35,5°, 38,9, JCPDS file No. 41-0254) is also detected in the sample with Cu:Ni (3:1)., i.e., alloyed.

The crystallite size of the copper, nickel and Cu-Ni alloy particles was calculated from the broadening of X-ray diffraction pattern using the Debey-Scherrer's equation (Equation 2).

$$D_{hkl} = \frac{K\lambda}{\beta_{2\theta}\cos\theta_B} \qquad (2)$$

Where D_{hkl} is the average crystallite size (nm), $\beta_{2\theta}$ is the broadening of the full width at half maximum (FWHM) of the most intense peak (111) in radians, θ_B is the Bragg's angle (degrees), K is the Scherrer's constant (a value of K=0.94 for spherical crystals with cubic symmetry is assumed in this work), and λ is the radiation wavelength (nm). This formula is applicable to crystal structures which particle size is smaller than 100 nm. The estimated values for FWHM and θ_B were obtained from fitting the diffraction pattern at (111) plane with a Lorentzian function [115]. The results are shown in Table 4.

Table 4. Results of crystallite size of selected samples Cu, Ni and Cu-Ni/AC

Sample	θ_B (degree)	FWHM (degree)	Crystallite size (nm)
Cu/AC	43.25	0.38063	23.5
Ni/AC	44.63	0.52013	17.3
Cu-Ni(2:1)/AC	43.57	0.37325	23.9
Cu-Ni(3:1)/AC	43.55	0.40036	22.2

From Table 4, the crystallite sizes of bimetallic samples Cu-Ni(2:1)/AC (23.9 nm) and Cu-Ni(3:1)/AC (22.2 nm) are closer to the size of Cu (23.5 nm) than to Ni (17.3 nm), which suggests that the copper structure would act as *hosting* for nickel atoms in the bimetallic system. Similar behavior is reported by Yang *et al.* [107] who showed that among several dopants for Cu-based catalyst Pd, Pt and Au prefer to stay in the surface of Cu(111), while Ni or Rh favor the bulk.

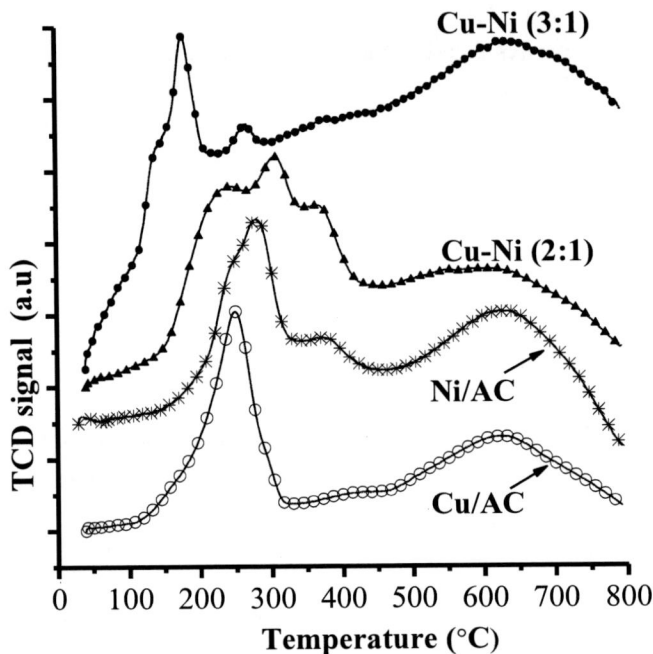

Figure 7. TPR profiles for calcined monometallic Cu/AC, Ni/AC and bimetallic Cu:Ni-3:1/AC and Cu-Ni/AC 2:1 samples.

4.2.3. Temperature Programmed Reduction

TPR profile of fresh Cu monometallic catalyst (Figure 7) exhibits a main reduction peak at 251°C and a weak shoulder at 290°C, attributed to a two-step reduction of CuO to Cu_2O and Cu^0, respectively. TPR of Ni monometallic catalyst sample shows a main reduction peak at 281°C and a smaller one at 380°C; both peaks are ascribed to reduction of NiO to Ni^0 [116]. The shoulder at 242°C is attributed to the reduction of well–dispersed NiO particles on AC support [84]. Additional peaks at 306°C in Cu-Ni:2-1 and 177°C in Cu-Ni:3-1, not found for the monometallic samples, were observed in the bimetallic samples. This would be another indication of the formation of Cu-Ni alloy species [84]. The broad peak observed after 500 °C, with a maximum at 600

°C, may be associated with the partial gasification of the activated carbon support as a result of CH_4 formation from the reaction of C and H_2 [117].

4.2.4. Scanning Electron Microscopy SEM

Figure 8 shows the SEM images of activated carbon, and Cu:Ni-3:1/AC and Cu:Ni-2:1/AC bimetallic catalysts.

Some elongated and cylindrical shapes can be noted in the catalysts, with the active metal particles homogeneously dispersed on the surface of AC. In addition, Cu:Ni-3:1/AC and Cu:Ni-2:1/AC bimetallic catalysts showed cluster agglomeration on the porous channels of AC support.

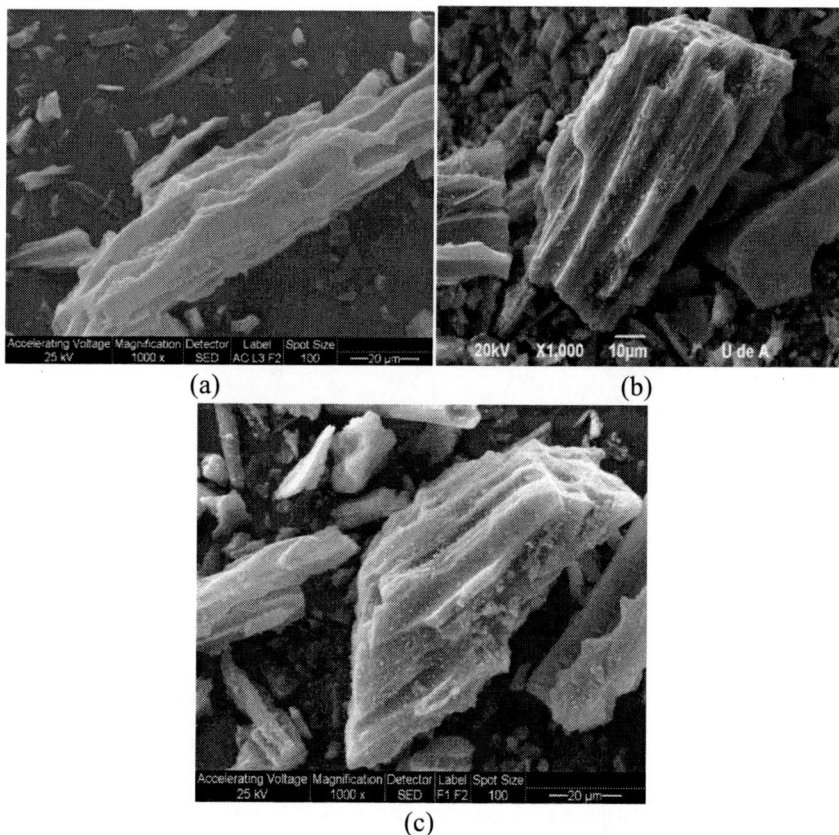

(a)   (b)

(c)

Figure 8. SEM images of (a) AC support, (b) Cu:Ni-3:1/AC, (c) Cu:Ni-2:1/AC.

5. DISCUSSION

Based on the results of characterization and catalytic activity presented above, it can be inferred that the activity of the catalyst is promoted not only due to a synergistic effect between copper and nickel (most bimetallic catalysts are more active than monometallic samples), but also to the formation of Cu-Ni alloy which would be the most active species in the reaction. The formation of the alloy likely occurs via mingling of copper and nickel at atomic scale, resulting in a regular crystalline phase. TPR profiles of the Cu-Ni alloy showed a degree of reducibility intermediate between the Cu and Ni metals, indicating the presence of three phases in the same catalyst. Considering that the monometallic catalysts, Cu and Ni, catalyze the reaction to a lesser extent, it is also expected a contribution of the individual metals in the catalytic performance of the sample containing the alloy phase Cu-Ni. In other bimetallic samples (Cu:Ni 1:3, 1:2, 1:2, 1:1), in which the presence of the alloy was not confirmed, Cu and Ni individual metals would catalyze the reaction independently, without having a significant synergistic effect. Results of N_2 adsorption, pulse chemisorption and TEM, show that Cu:Ni-2:1/AC and Cu:Ni-3:1/AC bimetallic catalysts, which reveal a higher catalytic activity, exhibit high surface area, very low metal dispersion and a higher mean particle size, which would indicate a strong effect of the metal particle size on catalyst activity.

CONCLUSION

The linear carbonates (DMC/DEC) were synthetized from CO_2 and alcohols (methanol/ethanol respectively) using supported bimetallic Cu-Ni catalyst as a safe and attractive methodology in terms of fixation of CO_2 and green chemistry.

Catalysts characterization of Cu:Ni-2:1/AC and Cu:Ni-3;1/AC samples revealed higher surface area than monometallic catalysts, formation of a Cu-Ni alloy, and co-existence of metal sites of Cu and Ni and species Cu^{2+} and Ni^{2+}. In addition, the metal particles were well dispersed on the surface of carbon; in addition, the crystallite size of the bimetallic catalysts would indicate that Cu structure can act as *hosting* for nickel atoms in the face-centred cubic phase of the alloy.

The synergistic interaction between Cu and Ni may explain the increased activity compared to copper and nickel mono-metallic samples. The significantly-improved catalytic activity of Cu-Ni/AC sample with Cu:Ni molar ratio of 3:1 (TOF = 126 h^{-1}) and 2:1 (TOF = 74 h^{-1}), compared to copper (TOF = 0.4 h^{-1}) and nickel (TOF = 0.2 h^{-1}) mono-metallic samples, seems to be correlated with the formation of Cu-Ni alloy species during the pyrolysis and activation stages.

ACKNOWLEDGMENTS

Financial support of the Universidad de Antioquia CODI Project Grant No.EO1557 and sustainability strategy 2013–2014 is gratefully acknowledged

REFERENCES

[1]   Schaffner, B.; Schaffner, F.; Verevkin, S.; Borner, A. Organic carbonates as solvents in synthesis and catalysis. *Chem. Rev.*; 2010, 110, 4554-4581

[2]   Aresta, M.; Dibenedetto, A. Utilisation of CO_2 as a chemical feedstock: opportunities and challenges. *Dalton Transactions,* 2007, 27, 2975–2992.

[3]   Parrish, J. ; Salvatore, R.; Woong, K. Perpestives on alkyl carbonates in organic synthesis. *Tetrahedrom,* 2000, 56, 8207-8237

[4]   Yoshio, O. Catalysis in the production and reactions of dimethyl carbonate, an environmentally benign building block. *Appl. Catal. A.,* 1997,155, 133-166.

[5]   Abbas-Alli, G.S.; Swaminathan, S. Organic carbonates. *Chem. Rev.,* 1996 ,96, 951-976.

[6]   Pacheco, M.A.; Marshall, C. L. Review of dimethyl carbonate (DMC) manufacture and its characteristics as a fuel additive. *Energy Fuels,* 1997, 11, 2–29.

[7]   Tundo, P.; Moraglio, G.; Trotta, F. Gas-liquid phase-transfer catalysis: A new continuous-flow method in organic synthesis. *Ind. Eng. Chem. Res.,* 1989, 28, 881-890.

[8] Zevenhovem, R.; Eloneva, S.; Teir, S. Chemical fixation of CO_2 in carbonates: Routes to valuable products and long term storage. *Catalysis Today*, 2006, 115, 73-79.

[9] Molzahn, D.; Jones, M.E.; Hartwell, G. E.; Puga, J. Production of dialkyl carbonates using copper catalysts, US Pat., 5387708 1995

[10] Roh, N.; Dunn, B.; Pugmire, R.; E. Eyring, E.; Meuzelaar, H. Production of diethyl carbonate from ethanol and carbon monoxide over a heterogeneous catalytic flow reactor. *Fuel Processing Technology*, 2003, 83, 27–38.

[11] Matsukazi, T. Novel method for dimethyl carbonate synthesis using methyl nitrite. *Stud. Surf. Sci. and Catalysis*, 2003, 145, 447-450

[12] Zhen, J.; YueHua, S.; Shu Hua, C. Novel synthesis of diethyl carbonate over palladium/MCM-41 catalysts. *Catal. Lett.* 69 2000, 69, 153–156

[13] Orrego, A.; Arbelaez, O.; Bustamante, F.; Villa, A. In situ FTIR study on the direct synthesis of DMC from CO_2 and etanol over Cu-Ni/AC. 15th International Congress of Catalysis. Munich Germany, 2012

[14] Arbelaez, O.; Orrego, A.; Bustamante,F.; Villa, A. Direct syntheis of diethyl carbonate from carbon dioxide and ethanol over Cu-Ni bimetallic catalysts. Top. Catal. 2012, 55, 668-672.

[15] Keller, N.; Rebmann, G.; Keller, V. Review: Catalysts, mechanisms and industrial processes for the dimethylcarbonate synthesis. J. Mol. Catal. A: Chem., 2010, 317, 1−18.

[16] Ballivet-Tkatchenko, D.; Jerphagnon, T.; Ligabue, R.; Plasseraud, L.; Poinsot, D. The role of distannoxanes in the synthesis of dimethyl carbonate from carbon dioxide. *Appl. Catal., A.*, 2003, 255, 93−99.

[17] Zhong, S.; Wang, J.; Xiao, X.; Li, H. Dimethyl carbonate synthesis from carbon dioxide and methanol over Ni−Cu/MoSiO-(VSiO) catalysts. *Stud. Surf. Sci. Catal.*, 2000, 130, 1565−1570.

[18] Bian, J.; Wei, X. W.; Wang, L.; Guan, Z. P. Graphene nanosheet as support of catalytically active metal particles in DMC synthesis. *Chin. Chem. Lett.*, 2011, 22, 57−60.

[19] Almusaiteer, K. Synthesis of dimethyl carbonate (DMC) from methanol and CO_2 over Rh-supported catalyst, *Catal. Commun.*, 2009, 10, 1127−1131.

[20] Bian, J.; Wei, X.; Jin, Y.; Wang, L.; Luan, D.; Guan, Z. Direct synthesis of dimethyl carbonate over activated carbon supported Cu−based catalysts., *Chem. Eng. J.*, 2010, 165, 686−692.

[21] Sakai, S.; Fujinami, T.; Yamada, T.; Furusawa, S. Reaction of organotin alkoxides with carbon disulfide, carbony sulfide or carbon dioxide.Nippon Kagaku Kaishi, 1975, 10, 1789-1794.

[22] Yamazaki, N.; Nakahama, S.; F. Higashi, F. Study on Chemical Reactions of Carbon dioxide, Rep. Asahi Glass Found. Ind. Technol 33 (1978) 31-45.

[23] Yamazaki, N.; Nakahama, S. Polymers Derived from Carbon Dioxide and Carbonates, *Ind. Eng. Chem., Prod. Res. Dev.*, 1979, 18, 249-252.

[24] Kizlink, J. Synthesis of Dimethyl Carbonate from Carbon Dioxide and Methanol in the Presence of Organotin Compounds. *Collect. Czech. Chem. Commun.*, 1993, 58, 1399-1402.

[25] Kizlink, J.; Pastucha, I. Preparation of Dimethyl Carbonate from Methanol and Carbon Dioxide in the Presence of Organotin Compounds. *Collect. Czech. Chem. Commun.*, 1994, 59 2116-2118.

[26] Kizlink, J.; Pastucha, I. Preparation of Dimethyl Carbonate from Methanol and Carbon Dioxide in the Presence of Sn(IV) and Ti(IV) Alkoxides and Metal Acetates. *Collect. Czech. Chem. Commun.*, 1995, 60, 687-692.

[27] Choi, J.C.; Sakakura, T.; Sako, T. Reaction of dialkyltin methoxide with carbon dioxide relevant to the mechanism of catalytic carbonate synthesis. *J. Am. Chem. Soc.* 1999, 121, 3793−3794.

[28] Ballivet-Tkatchenko, D.; Douteau, O.; Stutzmann, S. Reactivity of carbon dioxide with n-butyl(phenoxy)-, (alkoxy)-, and (oxo) stannanes: Insight into dimethyl carbonate synthesis *Organometallics*, 2000, 19, 4563−4567.

[29] Ballivet-Tkatchenko, D.; Jerphagnon, T.; Ligabue, R.; Plasseraud, L.; Poinsot, D. The role of distannoxanes in the synthesis of dimethyl carbonate from carbon dioxide. *Appl. Catal. A.,* 2003, 255, 93−99.

[30] Ballivet-Tkatchenko, D.; Ligabue, R. A.; Plasseraud, L. Synthesis of dimethyl carbonate in supercritical carbon dioxide. Braz. J. Chem. Eng, 2006, 23, 111−116.

[31] Ballivet-Tkatchenko, D.; Chambrey, S.; Keiski, R.; Ligabue, R.; Plasseraud, L.; Richard, P.; Turunen, H. Direct synthesis of dimethyl carbonate with supercritical carbon dioxide: Characterization of a key organotin oxide intermediate. *Catal. Today.*; 2006, 115 80−87.

[32] Kohno, K.; Choi, J.; Ohshima, Y.; Yili, A.; Yasuda, H.;Toshiyasu Sakakura,TReaction of dibutyltin oxide with methanol under CO_2 pressure relevant to catalytic dimethyl carbonate synthesis. *J. Organomet. Che*m., 2008, 693, 1389–1392.

[33] Plasseraud, L.; Ballivet-Tkatchenko, D.; Cattey, H.; Chambrey, S.; Ligabue, R.; Richard, P.; Willem, R.; Biesemans, M. Di-n-butyltin oxide as a chemical carbon dioxide capturer. *J. Organomet. Chem.*, 2010, 695, 1618−1626.

[34] Fan, B.; Zhang, J.; Li, R.; Fan, W. In situ preparation of functional heterogeneous organotin catalyst tethered on SBA-15. *Catal. Lett.*, 2008, 121, 297−302.

[35] Fan, B.; Li, H.; Fan, W.; Zhang, J.; Li, R. Organotin compounds immobilized on mesoporous silicas as heterogeneous catalysts for direct synthesis of dimethyl carbonate from methanol and carbon dioxide. *Appl. Catal. A.*, 2010, 372. 94−102.

[36] Fan, B.; Li, H.; Fan, W.; Qin, Z.; Li, R. Direct synthesis of dimethyl carbonate from methanol and carbon dioxide over organotin functionalized mesoporous benzene-silica. *Pure Appl. Chem.*, 2012, 84, 663−673.

[37] Sakakura, T.; Saito, Y.; Okano, M.; Choi, J.-C.; Sako, T. Selective conversion of carbon dioxide to dimethyl carbonate by molecular catalysis. *J. Org. Chem.*, 1998, 63 7095−7096.

[38] Isaacs, N. S.; O'sullivan, B.; Verhaelen, C. High pressure routes to dimethyl carbonate from supercritical carbon dioxide. *Tetrahedron*, 1999, 55, 11949−11956.

[39] Sakakura, T.; Choi, J.-C.; Saito, Y.; Masuda, T.; Sako, T.; Oriyama, T. Metal-catalyzed dimethyl carbonate synthesis from carbon dioxide and acetals. *J. Org. Chem.*, 1999, 64, 4506−4508.

[40] Sakakura, T.; Choi, J.-C.; Saito, Y.; Sako, T. Synthesis of dimethyl carbonate from carbon dioxide: catalysis and mechanism. *Polyhedron*, 2000, 19, 573−576.

[41] Choi, J.C.; He, L.N.; Yasuda, H.; Sakakura, T. Selective and high yield synthesis of dimethyl carbonate directly from carbon dioxide and methanol, *Green Chem.*, 2002, 4, 230−234.

[42] Zhao, T.; Han, Y.; Sun, Y. Supercritical synthesis of dimethyl carbonate from CO_2 and methanol. *Stud. Surf. Sci. Catal.*, 2000, 130, 461−466.

[43] Zhao, T.; Han, Y.; Sun, Y. Novel reaction route for dimethyl carbonate synthesis from CO_2 and methanol. *Fuel Process. Technol.*, 2000, 62, 187−194.

[44] Sun, Y. Chemicals from CO_2 via heterogeneous catalysis at moderate conditions. *Stud. Surf. Sci. Catal.*, 2004, 153, 9−16.

[45] Aresta, M.; Dibenedetto, A.; Fracchiolla, E.; Giannoccaro, P.; Pastore, C.; Pápai, I.; Schubert, G. Mechanism of formation of organic

carbonates from aliphatic alcohols and carbon dioxide under mild conditions promoted by carbodiimides. DFT calculation and experimental study. *J. Org. Chem.*, 2005, 70, 6177−6186.

[46] Dibenedetto, A.; Pastore, C.; Aresta, M. Direct carboxylation of alcohols to organic carbonates: Comparison of the group 5 element alkoxides catalytic activity: An insight into the reaction mechanism and its key steps. *Catal. Today,* 2006, 115, 88−94.

[47] Aresta, M.; Dibenedetto, A.; Pastore, C.; Pápai, I.; Schubert, G. Reaction mechanism of the direct carboxylation of methanol to dimethylcarbonate: Experimental and theoretical studies. *Top. Catal.,* 2006, 40, 71−81.

[48] Aresta, M.; Dibenedetto, A.; Nocito, F.; Pastore, C. Comparison of the behaviour of supported homogeneous catalysts in the synthesis of dimethyl carbonate from methanol and carbon dioxide: Polystyrene grafted tin−metallorganic species versus silesquioxanes linked Nbmethoxo species. *Inorg. Chim. Acta,* 2008, 361, 3215−3220.

[49] Aresta, M.; Dibenedetto, A.; Nocito, F.; Angelini, A.; Gabriele, B.; Negri, S. D. Synthesis and characterization of a novel polystyrene-tethered niobium methoxo species. Its application in the CO_2-based carboxylation of methanol to afford dimethyl carbonate.Appl. *Catal. A.,* 2010, 387, 113−118.

[50] Fang, S.; Fujimoto, K. Direct synthesis of dimethyl carbonate from carbon dioxide and methanol catalyzed by base. *Appl. Catal. A.*, 1996, 142, L1−L3.

[51] Cai, Q. H.; Lu, B.; Guo, L.; Shan Y. K. Studies on synthesis of dimethyl carbonate from methanol and carbon dioxide. *Catal. Commun.*, 2009, 10, 605–609.

[52] Cai, Q. H.; Jin, C.; Lu, B.; Tangbo, H.; Shana, Y. Synthesis of dimethyl carbonate from methanol and carbon dioxide using potassium methoxide as catalyst under mild conditions *Catal. Lett.*, 2005, 103, 225−228.

[53] Wang, H.; Lu, B.; Cai, Q. H.; Wu, F.; Shan, Y. K. Synthesis of dimethyl carbonate from methanol and carbon dioxide catalyzed by potassium hydroxide under mild conditions. *Chin. Chem. Lett.* 2005, 16, 1267−1270.

[54] Tomishige, K.; Ikeda, Y.; Sakaihori, T.; Fujimoto, K. A novel method of direct synthesis of dimethyl carbonate from methanol and carbon dioxide catalyzed by zirconia. *Catal. Lett.*, 1999, 58, 225−229.

[55] Tomishige, K.; Ikeda, Y.; Sakaihori, T.; Fujimoto, K. Catalytic properties and structure of zirconia catalysts for direct synthesis of

dimethyl carbonate from methanol and carbon dioxide. *J. Catal.*, 2000, 192, 355−362.

[56]   Tomishige, K.; Furusawa, Y.; Ikeda, Y.; Asadullah, M.; Fujimoto, K. CeO_2−ZrO_2 solid solution catalyst for selective synthesis of dimethyl carbonate from methanol and carbon dioxide. *Catal. Lett.*, 2001, 76, 71−74.

[57]   Tomishige, K.; Kunimori, K. Catalytic and direct synthesis of dimethyl carbonate starting from carbon dioxide using CeO_2−ZrO_2 solid solution heterogeneous catalyst: Effect of H_2O removal from the reaction system. *Appl. Catal. A.*, 2002, 237, 103−109.

[58]   Han, G. B.; Park, N. K.; Jun, J. H.; Chang, W. C.; Lee, B. G.; Ahn, B. S.; Ryu, S. O.; Lee, T. J. Synthesis of Dimethyl Carbonate from CH_3OH and CO_2 with $Ce_{1-x}Zr_xO_2$ Catalysts. *Stud. Surf. Sci. Catal.*, 2004, 153, 181−184.

[59]   Zhang, Z.F.; Liu, Z.T.; Liu, Z.W.; Lu, J. DMC formation over $Ce_{0.5}Zr_{0.5}O_2$ prepared by complex-decomposition method. *Catal. Lett.*, 2009, 129, 428−436.

[60]   Zhang, Z.F.; Liu, Z.W.; Lu, J.; Liu, Z.T. Synthesis of dimethyl carbonate from carbon dioxide and methanol over CexZr1−xO2 and [EMIM]Br/$Ce_{0.5}Zr_{0.5}O_2$. *Ind. Eng. Chem. Res.*, 2011, 50, 1981−1988.

[61]   Jiang, C.; Guo, Y.; Wang, C.; Hu, C.; Wu, Y.; Wang, E. Synthesis of dimethyl carbonate from methanol and carbon dioxide in the presence of polyoxometalates under mild conditions. *Appl. Catal. A.*, 2003, 256, 203−212.

[62]   Ikeda, Y.; Sakaihori, T.; Tomishige, K.; Fujimoto, K. Promoting effect of phosphoric acid on zirconia catalysts in selective synthesis of dimethyl carbonate from methanol and carbon dioxide. *Catal. Lett.*, 2000, 66, 59−62.

[63]   Ikeda, Y.; Asadullah, M.; Fujimoto, K.; Tomishige, K. Structure of the active sites on H_3PO_4/ZrO_2 catalysts for dimethyl carbonate synthesis from methanol and carbon dioxide. *J. Phys. Chem. B.*, 2001, 105, 10653−10658.

[64]   Eta, V.; Mäki-Arvela, P.; Leino, A.-R.; Kordás, K.; Salmi, T.; Murzin, D. Y.; Mikkola, J.P. Synthesis of dimethyl carbonate from methanol and carbon dioxide: Circumventing thermodynamic limitations. Ind. Eng. Chem. Res., 2010, 49, 9609−9617.

[65]   La, K. W.; Song, I. K. Direct synthesis of dimethyl carbonate from CH_3OH and CO_2 by $H_3PW_{12}O_{40}$/$Ce_xTi_{1-x}O_2$ catalyst *React. Kinet. Catal. Lett.*, 2006, 89, 303−309.

[66] La, K. W.; Jung, J. C.; Kim, H.; Baeck, S.-H.; Song, I. K. Effect of acid–base properties of H3PW12O40/CexTi1–xO2 catalysts on the direct synthesis of dimethyl carbonate from methanol and carbon dioxide: A TPD study of $H_3PW_{12}O_{40}/Ce_xTi_{1-x}O_2$ catalysts. *J. Mol. Catal. A: Chem.,* 2007, 269, 41–45.

[67] Lee, H. J.; Park, S.; Jung, J. C.; Song, I. K. Direct synthesis of dimethyl carbonate from methanol and carbon dioxide over $H_3PW_{12}O_{40}/Ce_xZr_{1-x}O_2$ catalysts: Effect of acidity of the catalysts., *Korean J. Chem. Eng.,* 2011, 28, 1518–1522.

[68] Yoshida, Y.; Arai, Y.; Kado, S.; Kunimori, K.; Tomishige, K. Direct synthesis of organic carbonates from the reaction of CO_2 with methanol and ethanol over CeO_2 catalysts *Catal. Today,* 2006, 115, 95–101.

[69] Aresta, M.; Dibenedetto, A.; Pastore, C.; Angelini, A.; Aresta, B.; Pápai, I. Influence of Al_2O_3 on the performance of CeO_2 used as catalyst in the direct carboxylation of methanol to dimethylcarbonate and the elucidation of the reaction mechanism. *J. Catal.,* 2010, 269, 44–52.

[70] Lee, H. J.; Park, S.; Song, I. K.; Jung, J. C. Direct Synthesis of Dimethyl Carbonate from Methanol and Carbon Dioxide over $Ga_2O_3/Ce_{0.6}Zr_{0.4}O_2$ Catalysts: Effect of Acidity and Basicity of the Catalysts. *Catal. Lett.,* 2011,141, 531–537.

[71] Lee, H. J.; Joe, W.; Song, I. K. Direct synthesis of dimethyl carbonate from methanol and carbon dioxide over transition metal oxide/$Ce_{0.6}Zr_{0.4}O_2$ catalysts: Effect of acidity and basicity of the catalysts. *Korean J. Chem. Eng.,* 2012, 29, 317–322.

[72] Eta, V.; Mäki-Arvela, P.; Salminen, E.; Salmi, T.; Murzin, D. Y.; Mikkola, J.P. The Effect of alkoxide ionic liquids on the synthesis of dimethyl carbonate from CO_2 and methanol over ZrO_2–MgO. *Catal. Lett.,* 2011, 141 1254–1261.

[73] Ballivet-Tkatchenko, D.; Dos Santos, J. H.; Philippot, K.; Vasireddy, S. Carbon dioxide conversion to dimethyl carbonate: The effect of silica as support for SnO_2 and ZrO_2 catalysts. *C. R. Chim.,* 2011, 14, 780–785.

[74] Allaoui, L.; Aouissi, A. Effect of the Bronsted acidity on the behavior of CO_2 methanol reaction. *J. Mol. Catal. A: Chem.* 2006, 259, 281–285.

[75] Honda, M.; Suzuki, A.; Noorjahan, B.; Fujimoto, K.; Suzuki, K.; Tomishige, K. Low pressure CO_2 to dimethyl carbonate by the reaction with methanol promoted by acetonitrile hydration. *Chem. Commun.,* 2009, 30, 4596–4598.

[76] Aouissi, A.; Apblett, A. W.; AL-Othman, Z. A.; Al-Amro, A. Direct synthesis of dimethyl carbonate from methanol and carbon dioxide using

heteropolyoxometalates: The effects of cation and addenda atoms. *Transition Met. Chem.*, 2010, 35, 927–931.

[77] Aouissi, A.; Al-Deyab, S. S.; Al-Owais, A.; Al-Amro, A. Reactivity of heteropolytungstate and heteropolymolybdate metal transition salts in the synthesis of dimethyl carbonate from methanol and CO_2. *J. Mol. Catal. A: Chem.*, 2010, 11, 2770–2779.

[78] Zhong, S.; Wang, J.; Xiao, X.; Li, H. Dimethyl carbonate synthesis from carbon dioxide and methanol over Ni–Cu/MoSiO-(VSiO) catalysts. *Stud. Surf. Sci. Catal.*, 2000, 130, 1565–1570.

[79] Li, C. F.; Zhong, S. H. Study on application of membrane reactor in direct synthesis DMC from CO_2 and CH_3OH over Cu–KF/MgSiO catalyst. *Catal. Today,* 2003, 82, 83–90.

[80] Wu, X.; Xiao, M.; Meng, Y.; Lu, Y. Direct synthesis of dimethyl carbonate on H_3PO_4 modified V_2O_5. *J. Mol. Catal. A: Chem.*, 2005, 238158–162.

[81] Wu, X.; Meng, Y.; Xiao, M.; Lu, Y. Direct synthesis of dimethyl carbonate (DMC) using Cu-Ni/VSO as catalyst. *J. Mol. Catal. A. Chem.*, 2006, 249, 93–97.

[82] Wang, X.; Xiao, M.; Wang, S.; Lu, Y.; Meng, Y. Direct synthesis of dimethyl carbonate from carbon dioxide and methanol using supported copper (Ni, V, O) catalyst with photo-assistance. *J. Mol. Catal. A. Chem.*, 2007, 278, 92–96.

[83] Bian, J.; Xiao, M.; Wang, S. J.; Lu, Y. X.; Meng, Y. Z. Highly effective direct synthesis of DMC from CH_3OH and CO_2 using novel Cu-Ni/C bimetallic composite catalysts. *Chin. Chem. Lett.,* 2009, 20, 352–355.

[84] Bian, J.; Xiao, M.; Wang, S.; Wang, X.; Lu, Y.; Meng, Y. Highly effective synthesis of dimethyl carbonate from methanol and carbon dioxide using a novel copper–nickel/graphite bimetallic nanocomposite catalyst. *Chem. Eng. J.*, 2009, 147, 287–296.

[85] Bian, J.; Xiao, M.; Wang, S.-J.; Lu, Y.-X.; Meng, Y.-Z. Carbon nanotubes supported Cu–Ni bimetallic catalysts and their properties for the direct synthesis of dimethyl carbonate from methanol and carbon dioxide. *Appl. Surf. Sci.*, 2009, 255, 7188–7196.

[86] Bian, J.; Xiao, M.; Wang, S.; Lu, Y.; Meng, Y. Direct synthesis of DMC from CH_3OH and CO_2 over V–doped Cu–Ni/AC catalysts. *Catal. Commun.*, 2009, 10, 1142–1145.

[87] Bian, J.; Xiao, M.; Wang, S.; Lu, Y.; Meng, Y. Novel application of thermally expanded graphite as the support of catalysts for direct

synthesis of DMC from CH_3OH and CO_2. *J. Colloid Interface Sci.*, 334 (2009) 50−57.

[88] Bian, J.; Xiao, M.; Wang, S.; Lu, Y.; Meng, Y. Graphite oxide as a novel host material of catalytically active Cu−Ni bimetallic nanoparticles. *Catal. Commun.*, 2009, 10, 1529−1533.

[89] Bian, J.; Wei, X. W.; Wang, L.; Guan, Z. P. Graphene nanosheet as support of catalytically active metal particles in DMC synthesis. *Chin. Chem. Lett.*, 2011, 22, 57−60.

[90] Almusaiteer, K. Synthesis of dimethyl carbonate (DMC) from methanol and CO_2 over Rh-supported catalyst. Catal. Commun. 2009, 10, 1127−1131.

[91] Bian, J.; Wei, X.; Jin, Y.; Wang, L.; Luan, D.; Guan, Z. Direct synthesis of dimethyl carbonate over activated carbon supported Cu−based catalysts. *Chem. Eng. J.*, 2010, 165, 686−692.

[92] Aouissi, A.; Al-Othman, Z. A.; Al-Amro, A. Gas-phase synthesis of dimethyl carbonate from methanol and carbon dioxide over $Co_{1.5}PW_{12}O_{40}$ Keggin-type heteropolyanion. *Int. J. Mol. Sci.*, 2010, 11, 1343−1351.

[93] Chen, Y.; Xiao, M.; Wang, S.; Han, D.; Lu, Y.; Meng, Y. Porous diatomite−immobilized Cu−Ni bimetallic nanocatalysts for direct synthesis of dimethyl carbonate. *J. Nanomater.*, 2012, 2012, 1−8.

[94] Chen, H.; Wang S.; Xiao M.; Han D.; Lu Y.; Meng Y. Direct Synthesis of Dimethyl Carbonate from CO_2 and CH_3OH using 0.4 nm Molecular Sieve Supported Cu-Ni Bimetal Catalyst. *Chin. J. Chem. Eng.* 2012, 20, 906−913.

[95] Bustamante, F.; Orrego, A.; Villegas, S.; Villa, A. Modeling of Chemical Equilibrium and Gas Phase Behavior for the Direct Synthesis of Dimethyl Carbonate from CO_2 and Methanol. *Ind. Eng. Chem. Res.*, 2012, 51, 8945–8956.

[96] Fujita, S.; Bhanage, B.; Ikushima, Y.; Arai, M. Synthesis of dimethtyl carbonate from carbon dioxide and methanol in the presence of methyl iodide and base catalysts under mild conditions. *Green Chem.*, 2001, 3 87-91

[97] Gasc, F.; Thiebaud-Roux, S.; Mouloungui, Z. Methods for sinthesizin diethyl carbonate from ethanol and super critical CO_2 by one or two steps in the presence of potassium carbonate. *J. Supercrit. Fluid.*, 2009, 50, 46-53

[98] Wang,W.; Wang,S.; Ma,X. Gong, J. Crystal structures, acid-base properties and reactivities fof $Ce_xZr_{1-x}O_2$ catalysts. *Catal. Today,* 2009,148, 323-328.

[99] Meng, Y. Z.: Bian, J.; Xiao, M. et al. "Catalyst used for catalytic synthesis for dimethyl carbonate directly from methanol and carbon dioxide, and preparation and using method thereof". CN Patent 101143322, 2008.

[100] Kim, K. J.; Lynch D. W. Electronic structure of Ni-Cu alloys studied by spectroscopic ellipsometry. Physical Review B., 1989, 39, 9882-9887.

[101] Baker, H.; Okamoto, H. ASM Handbook Volume 3: Alloy Phase Diagrams. ASM International. Ohio, 1992. ISBN: 978-0-87170-381-1

[102] Osaki, T.; Horiuchi, T.; Suzuki, K.; Mori, T. Catalyst performance of MoS_2 and WS_2 for the CO_2-reforming of CH_4. Suppression of carbon deposition. *Appl. Catal. A.* 1997, 155, 229-238.

[103] Kroll, V.; Tjatjopoulos, C.H.; Mirodatos, G.J. Kinetics of Methane Reforming over Ni/SiO_2 Catalysts ased on a Stepwise Mechanistic Model. *Stud. Surf. Sci. Catal.,* 1998, 119, 753-758.

[104] Ding, X.; Pagan, V.; Peressi, M.; Ancilotto, F. Modeling adsorption of CO_2 on Ni(110) surface. Mater. Sci. Eng., C, 2007, 27, 1355–1359.

[105] Ding, X.; De Rogatis, L.; Vesselli, E.; Baraldi, A.; Comelli, G.; Rosei, R.; Savio, L.; Vattuone, L.; Rocca, M.; Fornasiero, P.; Ancilotto, F.; Baldereschi, A.; Peressi, M. Interaction of carbon dioxide with Ni(110): A combined experimental and theoretical study. *Phys. Rev. B: Condens. Matter.,* 2007, 76, 1198-200.

[106] Lapidus, A. L.; Gaidai, N. A.; Nekrasov, N. V.; Tishkova, L. A.; Agafonov, Yu. A.; Myshenkova, T. N. The Mechanism of Carbon Dioxide Hydrogenation on Copper and Nickel *Catalysts. Petrol. Chem.,* 2007, 47, 75–82.

[107] Yang, Y.; White, M.G.; Liu, P. Theoretical Study of Methanol Synthesis from CO_2 Hydrogenation on Metal-Doped Cu(111) Surfaces. *J. Phys. Chem. C.,* 2012,116, 248–256.

[108] Nerlov, J.; Chorkendorff, I. Promotion through gas phase induced surface segregation: methanol synthesis from CO, CO_2 and H_2 over Ni/Cu(100). *Catal. Lett.,* 1998, 54, 171–176.

[109] Nerlov J.; Sckerl S.; Wambach J.; I. Chorkendorff. Methanol synthesis from CO_2, CO and H_2 over Cu(100) and Cu(100) modified by Ni and Co. *Appl. Catal. A. Gen.* 2000, 191, 97–109.

[110] Gan, L.Y.; Tian, R.-Y.; Yang, X.B.; Lu, H.D.; Zhao, Y.J. Catalytic Reactivity of CuNi Alloys toward H_2O and CO Dissociation for an

Efficient Water-Gas Shift: A DFT Study, J. *Phys. Chem. C.*, 2012, 116, 745–752.

[111] Naghash, A. R.; Etsell, T. H.; Xu, S. XRD and XPS Study of Cu-Ni Interactions on Reduced Copper-Nickel-Aluminum Oxide Solid Solution Catalysts. *Chem. Mater.* 2006, 18, 2480-2488.

[112] Mohamed, G. H.; Radiman, S.; Gasaymeh, S. S.; Lim, H. N.; Huang, N. M. "Mild Hydrothermal Synthesis of Ni–Cu Nanoparticles *J. Nano Mat.*, 2010, 2010,1-5

[113] Baskaran, I.; Sankara Narayanan, T. S. N.; Stephen, A. Pulsed electrodeposition of nanocrystalline Cu–Ni alloy films and evaluation of their characteristic properties. *Mater. Lett.*, 2006, 60, 1990–1995.

[114] Jung, C.H.; Lee, H.G.; Kim, C.J.; Bhaduri, S. B. Synthesis of Cu–Ni alloy powder directly from metal salts solution. *J. Nanopart. Res.*, 2003, 5, 383–388.

[115] Abdullah, M.; Khairurrijal. Derivation of Scherrer Relation Using an Approach in Basic Physics Course. *J. Nano Saintek .*, 2008, 1, 28-32

[116] Silva, L. M. S.; Órfão, J. J. M.; Figueiredo, J. L. Formation of two metal phases in the preparation of activated carbon-supported nickel catalysts,. *App. Catal. A. Gen.*, 2001, 209, 145-154.

[117] Nishiyama, Y.; Haga, T. Low Temperature Hydrogasification of carbons using nickel-base catalyst. *Stud. Surf. Sci. Catal.,* 1981, 7, 1434-1435.

In: Carbonates   ISBN: 978-1-62948-178-4
Editors: B.A. Hughes, T.C. Wagner   © 2013 Nova Science Publishers, Inc.

Chapter 2

FUNCTIONAL POLYMERS BASED ON CARBONATES OBTAINED FROM CO$_2$

*Bungo Ochiai[1] and Takeshi Endo[2]**

[1]Department of Chemistry and Chemical Engineering,
Faculty of Engineering, Yamagata University, Yamagata, Japan
[2]Molecular Engineering Institute, Kinki University, Fukuoka, Japan

ABSTRACT

This chapter describes synthesis and application of various functional polymers from carbonates obtained by the reaction of epoxides and CO$_2$. Main subjects are polyhydroxyurethanes prepared via polyaddition of cyclic carbonates and amines, and polymers bearing cyclic carbonate moieties prepared from glycidyl methacrylate (GMA) and CO$_2$.Polyhydroxyurethanes, polyurethanes bearing hydroxyl groups in the side chains, are prepared via polyaddition of bifunctional carbonates with diamines. This polyaddition is chemo-selective, and various functional monomers and polymerization conditions are available even though the primary reaction is a nucleophilic addition. Transformation of hydroxyl side chains also introduces various functional groups. The potential applications involve paints, packaging materials, biomaterials, and electronic devices. Polymers bearing carbonate moieties in the side chain are potentially applicable as optical materials, polymeric electrolytes, adhesives, and paints. One of the most facile methods is the reaction with

---

* Corresponding author: E-mail: tendo@moleng.fuk.kindai.ac.jp.

GMA and CO_2, in which three pass ways are possible.The synthetic methods are polymerization of a methacrylate bearing carbonate moieties obtained by the reaction of GMA and CO_2, reaction of polyGMA and CO_2, and concurrent polymerization and CO_2 fixation via radical polymerization of GMA under CO_2 atmosphere.

ABBREVIATIONS

AIBN	2,2'-azobisisobutyronitrile
BisAC	2,2-bis(1,3-dioxolane-2-one-4-ylmethoxyphenyl)propane
BMI	1-butyl-3-methylimidazolium
CL	ε–caprolactone
CP MAS	cross polarization magic angle scanning
DBU	1,8-diazabicyclo[5.4.0.]undec-7-ene
DBN	1,5-diazabicyclo[4.3.0]nona-5-ene
DETA	diethylenetriamine
DMAP	N,N-dimethylaminopyridine
DMF	N,N-dimethylformamide
DOA	(1,3-dioxoran-2-one)ylmethyl acrylate
DOMA	(1,3-dioxoran-2-one)ylmethyl methacrylate
EC	ethylene carbonate
EO	ethylene oxide
GA	glycidyl acrylate
GMA	glycidyl methacrylate
M_n	number–average molecular weight
M_w	weight–average molecular weight
NMR	nuclear magnetic resonance
PC	propylene carbonate
PO	propylene oxide
salen	salicylaldimine
TMU	tetramethylene urea
T_g	glass transition temperature

1. INTRODUCTION

CO_2 is a non-petroleum resource exists in the atmosphere at the concentration around 400 ppm. Although this concentration is rather low, the amount may be calculated to be in the order of Tt, namely 10^{12} t, by the very large amount of the atmosphere. This amount is obviously larger than the amounts of plastic production, ca. 2×10^8 t/y, and we may estimate that the industrial consumption of CO_2 negligibly affects the needs of CO_2 for photosynthesis by plants. In the viewpoint as an industrial resource, the low toxicity and the inexpensiveness are also the advantages. Moreover, it is preferable if CO_2, which is a major greenhouse gas emitted by various industrial and biological activities, can be reduced by transformation into useful materials. We may obtain gas with higher CO_2 concentration and higher temperature as exhausts of industrial processes and thermal power stations.

These situations push CO_2 to be an attractive carbon source replacing limited petroleum resources. However, CO_2 is the final stable product of various combustions requiring high energy for its transformation. As a result, industrial processes converting CO_2 are very limited. Major transformations are urea synthesis from NH_3 and CO_2 and carbonate synthesis from epoxides and CO_2. Recently, EC obtained from EO and CO_2 was applied for industrial synthesis of polycarbonate, excluding toxic and corrosive phosgene derivatives [1]. Other various polymers have also been obtained using CO_2, and development of a wide variety of functional polymers based on CO_2 will extend the utility of CO_2 as a useful resource. Because many polymerizations utilizing CO_2 have been described in several excellent reviews [1-12], this chapter focuses on the development of functional polymers based on cyclic carbonates prepared from CO_2 and epoxides.

2. SYNTHESIS OF CYCLIC CARBONATES FROM CARBON DIOXIDE

The reaction of CO_2 with epoxides is the most examined reaction for CO_2 (Scheme 1). Various metal, ammonium, and phosphonium compounds catalyze the reaction of various epoxides. This reaction may proceed under mild conditions (e.g., atmospheric pressure) and affords five-membered cyclic carbonates in high yields. Transition metal catalysts are typically used for epoxides with lower reactivity such as EO and PO. This industrialized reaction is conducted under high pressure (e.g., > 10 MPa) and high temperature (e.g., > 100 °C). The reactions of epoxides such as epichlorohydrin and glycidyl ethers proceed with simple alkaline metal or ammonium halides under milder conditions, e.g., 80 °C under an atmospheric pressure. In the case of group 1 metal catalysts, the activities in the reaction at atmospheric pressure basically depend on the Lewis acidities of the cations and the nucleophilicities of the anions, whereas solubilities of catalysts sometimes affect the activities [13]. Recently, amidines were proved as effective carriers for CO_2. *N*-Methyl tetrahydropyrimidine accelerates the carbonation of various epoxide catalyzed by LiBr. As a result, the reaction proceeds under very mild consitions (e.g., 45 °C and 1 atm) [14,15]. Hydrogen iodide salts of commercial bicyclic amidines (DBN and DBU) are very effective catalysts for carbonation of epoxy resins, and the reaction proceeds within 24 h even at room temperature under 1 atm of CO_2 atmosphere [16,17].

Two reaction mechanisms were proposed for this reaction as shown in Scheme 1. The reaction mechanisms are postulated to depend on the pressure of CO_2. Under low pressure, the epoxide ring-opens with a catalyst to form an alkoxide, and the subsequent nucleophilic addition to CO_2 followed by cyclization gives a cyclic carbonate. Under high pressure, the nucleophilic addition of a catalyst to CO_2 priory proceeds to give a carboxylate anion. Then, the epoxide ring-opens by the nucleophilic addition of the anion, and the subsequent cyclization gives a cyclic carbonate. An earlier review by Darensbourg and Holtcamp has described the reaction behavior in detail [8]. Polymers bearing quaternary ammonium salts moieties and modified ion-exchange resins were also employed as catalysts to attain easy separation and recycle of catalysts [18-21].

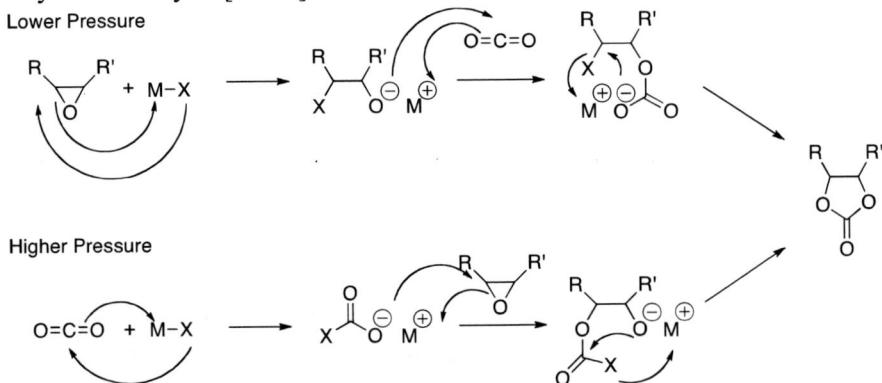

Scheme 1.

Five-membered cyclic carbonates with vinyl [22,23] and exomethylene [24-27] structure can also be prepared from CO_2. The radical polymerization of the unsaturated carbonates gives polymers with pendant cyclic carbonate moieties [23,24,27,28-30]. GA [31-33] and GMA [34,35] may be transformed to the corresponding (meth)acrylates bearing a cyclic moiety, and their polymerizations also give polymers with pendant cyclic carbonate moieties (discussed later with incorporation of CO_2 into polymers bearing epoxy moieties).

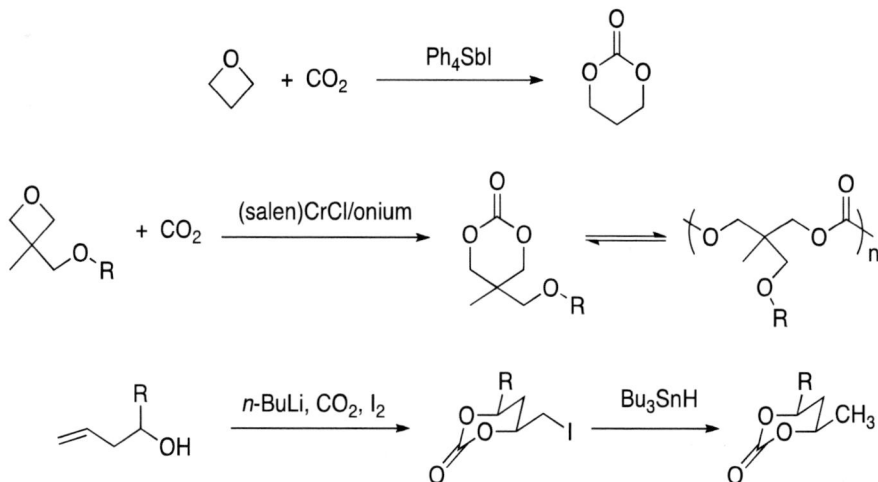

Scheme 2.

Reaction of CO_2 with oxetane provides a six-membered cyclic carbonate, 1,3-dioxolan-2-one, using Ph_4SbI as a catalyst (Scheme 2) [36]. However, reactions of substituted oxetanes were not reported in this literature. The reaction of 3-alkoxymethyl-3-methyl oxetanes and CO_2 proceeds by a catalyst system consisting of (salen)CrCl and onium salts [37]. The produced 5-alkoxymethyl-5-methyl-1,3-dioxolan-2-ones are in equilibria between the polymers produced via ring-opening polymerizations of the cyclic carbonates initiated by the catalyst.

As another approach to obtain six-membered cycliccarbonates from CO_2, homoallyl alcohols were reacted with CO_2 and iodine stereo-specifically in the presence of n-butyllithium to yield cyclic carbonates bearing iodomethyl groups that may be reduced to methyl groups by treatment with tributyltin hydride [38-40].

Analogous five-membered cyclic carbonates may be prepared in a similar manner by employing allyl alcohols [38-40].

3. STEP-POLYMERIZATIONOF CYCLIC CARBONATES

Aminolysis of cyclic carbonates gives urethanes with hydroxyl groups (hydroxyurethanes) [30,34,41-83]. Primary and cyclic secondary amines are employed for the reaction, but sterically hindered amines are not able to react with cyclic carbonates. Because hydroxyurethanes produced in the reaction with five-membered cyclic carbonates may contain primary and secondary alcohol moieties, depending on which of C–O linkages breaks, this reaction usually gives mixtures of two isomers, in which hydroxyurethanes with secondary alcohol moieties are the major products. The content of the secondary alcohol moieties increases on introduction of an electron-withdrawing group to the α-position of the cyclic carbonate, and the trend was supported by the *ab initio* calculation [44,45]. The effect of ring-size on aminolysis behavior was compared for five-, six-, and seven-membered cyclic carbonates, revealing that larger ring-size cyclic carbonates aminolyzed faster by the larger ring strains [46,47].

Polyaddition of bifunctional cyclic carbonates with diamines yields polyurethanes with hydroxyl groups (i.e., polyhydroxyurethanes) (Scheme 3) [48-71]. Five-membered cyclic carbonates have been the most focused carbonate monomers due to their facile preparation methods and the wide variety of the accessible precursors. Accordingly, this chapter describes the polymerization employing five-membered cyclic carbonates. Polymerizations of larger-ring carbonates have been described in a previous review [10]. The advantages of this polyaddition are the chemo-selectivity and no need of toxic and unstableisocyanates required for typical polyurethane synthesis. The excellent chemo-selectivity resulted in the high tolerance to various reaction conditions such as air and water, and the wide variety of functional groups introducable to the polymers. Cyclic carbonate monomers may be prepared *in situ* from diepoxides and CO_2, and subsequent one-pot addition of diamines successfully gave polyhydroxyurethanes having identical structures with those prepared from purified monomers [64].

As a result of the excellent chemo-selectivity, molecular weights and yields were negligibly changed by the addition of water, alcohols, or an ester even though the primary reaction is nucleophilic [62]. Polyaddition in aqueous dispersions is also possible with neither organic solvents nor surfactants. When dispersions of hydrophobic carbonate monomers in aqueous solution of diamines were stirred at 60°C, the heterogeneous polyaddition proceeded prior to the hydrolysis of the carbonate ring to yield polyhydroxyurethanes in excellent yields. The isolation process is simple filtration and washing. The polyaddition at higher temperatures and the use of hydrophilic carbonates were unsuccessful due to the hydrolysis of the carbonate moieties. Polyaddition in ionic liquids also proceeded, and the available monomers and the reaction conditions are identical to the conventional polymerizations employing organic solvents [67]. The products are ionic composites *in situ* prepared during polyaddition by interaction of hydroxyl groups with the ionic liquid (Figure 1a). This polyaddition may also be conducted in a mixed solvent of [BMI][PF$_6$] and water, and the produced polyhydroxyurethane was deposited from the solvent (Figure 1b). Both polyhydroxyurethane and [BMI][PF$_6$] could be isolated by simple filtration and pipetting, respectively.

The chemo-selective nature also allows the use of various functional monomers, which cannot be applied to conventional polyurethane synthesis employing highly reactive isocyanates. L-Lysine containing carboxylic acid could be polymerized in the presence of DBU or sodium hydroxide to neutralize the carboxyl group, although the yields were relatively low [55]. An attempt to increase the yield by protecting the carboxyl moieties by esterification was not successful, probably because of the poor nucleophilicity of the α-amino group owing to the electron-withdrawing ester. By contrast, the use of L-lysinol prepared by reduction of L-lysine gave the corresponding polyhydroxyurethane in a quantitative yield. The resulting polyhydroxyurethanes are optically active and the crosslinking reaction with metal salts such as cupric acetate gave chiral gels that are potentially applicable as chiral stationary phase and chiral reagents. DETA bearing two primary and one secondary amine structures may also be used as a monomer for selective polyaddition with the primary amine structure remaining the secondary amine structure [63]. The resulting polyhydroxyurethane may be crosslinked via the reactions of the secondary amine structure. The polyaddition of a five-membered cyclic carbonate bearing chloromethyl moieties and DETA gave a branched cationic polyhydroxyurethane (Scheme 4) [68]. First, the primary amine structure selectively reacts with the carbonate ring, and then the remained chloromethyl group quaternarizes the secondary amine group to form a quaternary ammonium structure. The resulting cationic polymer serves as a recyclable catalyst for the reaction of epoxides with CO_2. Owing to the cationic structure, this polymer forms stable complexes with DNA. The cytotoxicity of the complex and the polymer are low, and this complex is potentially applicable to gene delivery.

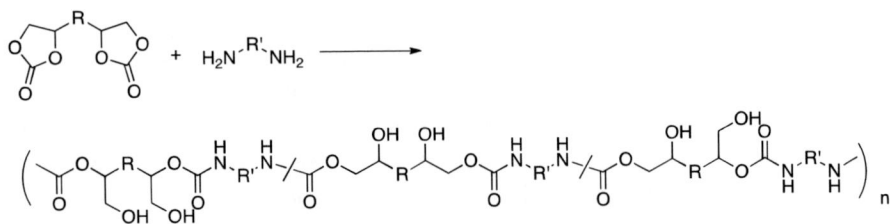

Scheme 3.

a

b

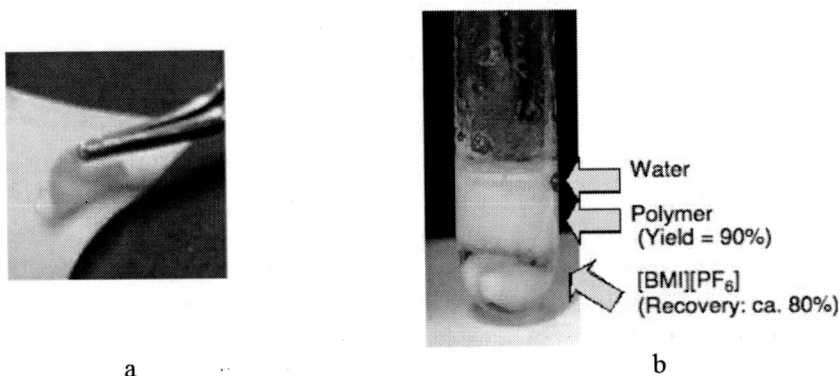

Figure 1. Photo images of (a) the ionic composite *in situ* prepared by polyaddition in [BMI][PF$_6$] and (b) the polymerization mixture after polyaddition in a mixed solvent of [BMI][PF$_6$]/water (v/v =2/1).

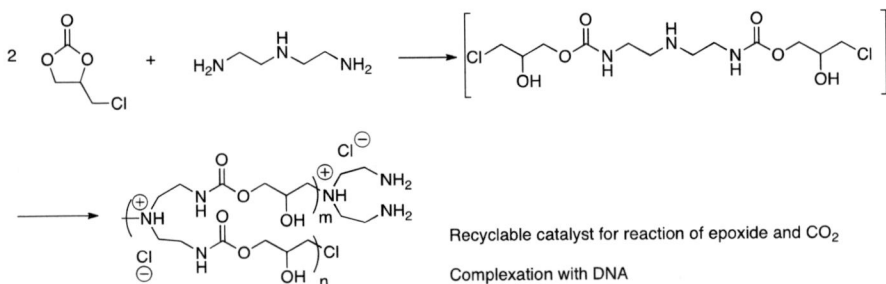

Recyclable catalyst for reaction of epoxide and CO$_2$

Complexation with DNA

Scheme 4.

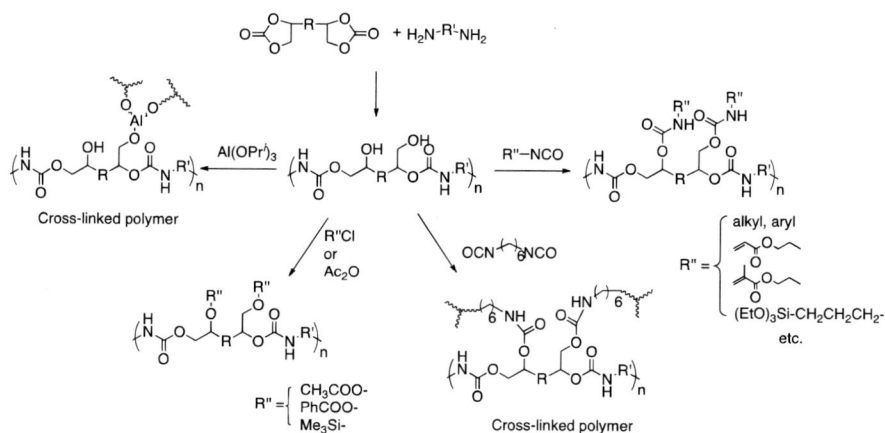

Scheme 5.

Various functional groups may be introduced via the reaction of the hydroxyl side chains (Scheme 5). After the aforementioned one-pot polymerization from diepoxides, CO_2, and diamines, the hydroxyl side chains may be transformed via subsequent one-pot reactions [64]. The treatments with acetic anhydride, benzoyl chloride, and trimethylsilyl chloride in the presence of DMAP gave the corresponding polyurethanes bearing ester or ether moieties in the side chains with excellent transformation efficiency. The transformation with various isocyanates was also examined [66,67]. Simple phenyl and ethyl isocyanates gave corresponding polyurethanes bearing urethane side chains [67]. The reaction of 2-methacryloyloxyethyl methacrylate gave polyurethane bearing radically and thermally cross-linkable polyurethane, in which the methacrylate contents may be controlled by the feed ratio [66]. The reaction of 3-triethoxysilylpropyl isocyanate gave polyurethane bearing triethoxysilyl groups in the side chain [66]. Sol-gel reaction of this reactive polyurethane with tetraethoxysilane gave polymer/silica composites. As other crosslinking reactions, reactions with hexamethylenediisocyanate and $Al(OPr^i)_3$ were also reported [54]. The crosslinking with $Al(OPr^i)_3$ proceeded very rapidly, and the gelation was completed within 1 min.

Table 1 summarizes the T_g values of polyhydroxyurethane prepared from BisAC and dodecandiamine and its various derivatives [64,67]. The T_g of the original polyhydroxyurethane is 46 °C. The conversions of the hydroxyl groups into ester and silyl ether groups decrease the glass transition temperatures by the loss of intermolecular hydrogen bonding. By contrast, the conversion to urethane structures increases the glass transition temperatures. IR and ^{1}H-NMR spectroscopic analyses suggested the presence of intermolecular hydrogen bonding between the urethane structures in both main and side chains. Namely, the urethane structures in the side chains presumably increase the entanglement of the polymer chains.

In spite of the polyurethane backbone of polyhydroxyurethanes, some properties of polyhydroxyurethanes differ with those of conventional polyurethanes. Polyhydroxyurethanes are amorphous without crystalline states, whereas typical polyurethanes are crystalline. An important factor is the presence of the side chains in polyhydroxyurethanes to prevent stacking of the main chain. The randomly existing units containing primary and secondary hydroxyl groups are also a reason for the amorphous nature. The authors' group has compared the properties of a polyhydroxyurethane obtained by a bifunctional cyclic carbonate derived from bisphenol-A diglycidyl ether (BisAC) and 1,6-hexanediamine with those of the analogous polyurethane obtained by polyaddition of bisphenol-A bis(2-hydroxyethyl ether) and hexamethylenediisocyanate [54]. The slightly higher T_g of the polyhydroxyurethane than that of the polyurethane (ΔT_g = 10 K) is attributed to hydrogen bonding between hydroxyl groups. The presence of hydroxyl groups also influences on the solubility of the polyhydroxyurethane. That is, this polyhydroxyurethane is almost insoluble in less polar solvents dissolving the polyurethane such as dichloromethane, chloroform, and acetone. On the other hand, the thermogravimetric change behaviors are almost identical.

Crosslinking polyadditions using trifunctional cyclic carbonates were also demonstrated [53,61]. Modification of epoxy resins with trifunctional cyclic carbonates effectively suppressed the exothermic effect and the increase of viscosity [53], and a similar behavior was also observed in the curing of epoxy resins modified by CO_2 [50,51]. These results suggested potential application as an active thinner for epoxy resins [53]. The effect of amines on thermal behaviors of the resulting gel was also investigated [61]. Monofunctional cyclic carbonates are also applicable as modifiers for epoxy resins reducing fragility of cured epoxy resins [69]. This report described the acceleration effect of protic solvents in the aminolysis of cyclic carbonates. Polyhydroxyurethane synthesis from natural raw material was reported [60,70]. For example, reaction of epoxydized soybean oil with CO_2 gave carbonated soybean oil in the presence of an ammonium catalyst [60]. Polyaddition of the soybean oil carbonate with di- or triamines gave the corresponding crosslinkedpolyhydroxyurethanes.

Table 1. Glass transition temperatures of polyhydroxyurethane obtained from BisAC and dodecanediamine and its derivatives

Reactant	Original polyhydroxy-urethane	Modified polyhydroxyurethanes				
		(MeCO)$_2$O	PhCOCl	Me$_3$SiCl	PhNCO	EtNCO
Side chain functions	HO-	MeCOO-	PhCOO-	Me$_3$SiO-	PhNHCOO-	EtNH-COO-
M_n (M_w/M_n)[a]	26200 (1.71)	9200 (1.75)	7200 (1.39)	7700 (1.63)	25700 (2.36)	21900 (2.16)
T_g (°C)[b]	46	26	17	14	74	53

[a] Estimated by SEC (DMF containing 10 mMLiBr, polystyrene standard).
[b] Determined by DSC (scaning rate = 10°C, N$_2$, second heating scan).

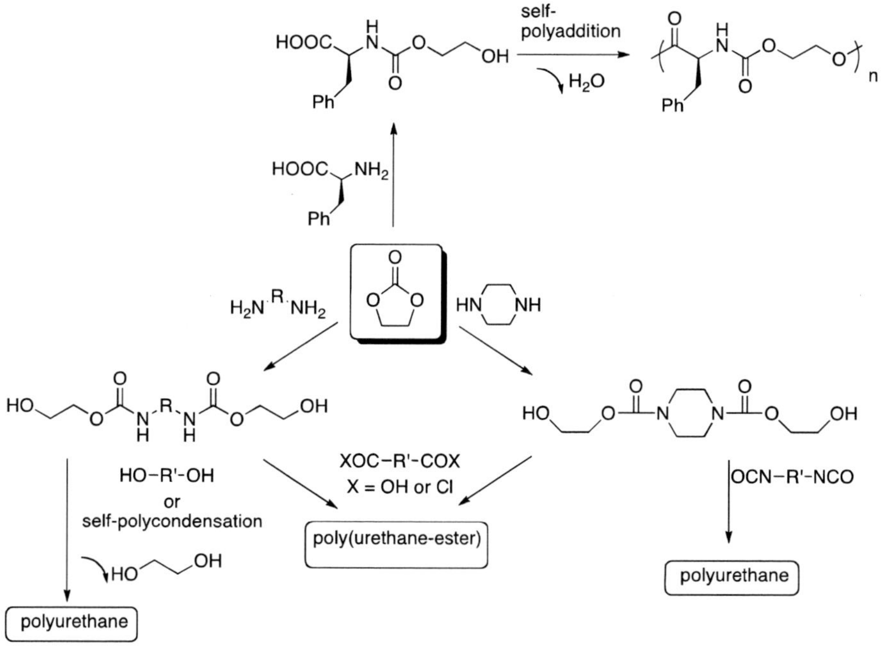

Scheme 6.

Both EC and PC were employed for synthesis of poly(urethane-ester)s and polyurethanes (Scheme 6). A chiral urethane with α,ω-hydroxycarboxylic acid structure was synthesized from L-phenylalanine and EC in the presence of bases to neutralize the carboxylic acid. Its self-polycondensation afforded a chiral poly(urethane-ester) [72]. Hydrolytic digestion of the chiral polymer, which could be accelerated by trypsin, was demonstrated. Other poly(urethane-ester)s were also obtained from urethane diols obtained by reactions of EC and PC with various aliphatic diamines. A urethane diol obtained by reaction of piperadine and EC was employed as a monomer for synthesis of poly(urethane-ester)s and alternating polyurethane copolymers by polycondensations with dicarboxylic acid dichloride and by polyaddition with diisocyanates, respectively [73]. The carbamate groups in the urethane diols contained no NH moieties, which cannot be obtained by conventional polyaddition of diols and diisocyanates. As another method to obtain poly(urethane-ester)s, ternary polycondensation of the urethane diols with aliphatic diols and adipic acids was conducted using supported *Candida antarctica* lipase B as a catalyst [74]. Adipic acid must be added in portions to avoid inhibition of the enzyme, and water formed during reaction should be removed under reduced pressure.

Rokicki and Piotrowska applied urethane diols obtained from EC and aliphatic diamines to synthesis of aliphatic polyurethanes [75]. Polycondensation of the urethane diols with aliphatic diols proceeded with elimination of ethylene glycol (i.e., transurethanization polymerization) in the presence of Bu_2SnO as a catalyst to afford polyurethanes, although the molecular weights are relatively lower (M_n< 4000). Urethane diols prepared from aminoalcohols and EC can also be used as monomers. In the self-polycondensation of the urethane diol prepared from 1,6-hexanediamine and EC, two-step polycondensation, consisting of the polycondensation under a N_2 atmosphere followed by that under reduced pressure, was effective to improve the yield and the molecular weight up to 90% and 10,000, respectively [76].Although the second polycondensation step at 180 °C was accompanied by formation of urea groups, this side reaction was relatively suppressed at 150 °C. The resulting polyurethane is telechelic polyurethane having hydroxyl groups at both of the end groups, and was converted to polyurethane methacrylate via a reaction with GMA. The resulting polyurethane methacrylate is thermally crosslinkable and served as a crosslinker for radical polymerization of methyl acrylate (Scheme 7).

Some step polymerizations employing five-membered cyclic carbonates have also been reported (Scheme 8) [77-83]. These polymerizations have already been described in earlier reviews and are not described herein [10].

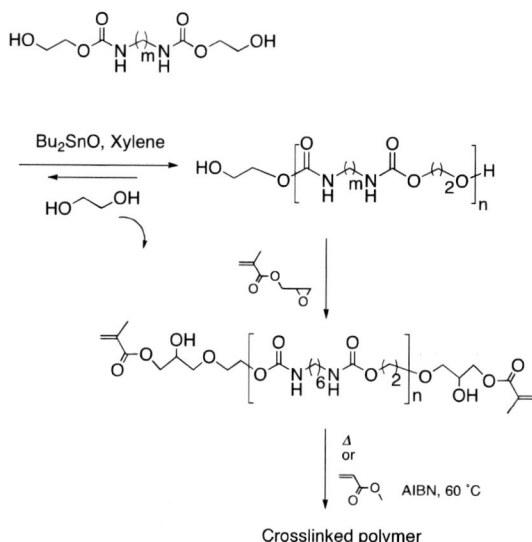

Scheme 7.

4. RING-OPENING POLYMERIZATION OF CYCLIC CARBONATES

Thermodynamically stable five-membered cyclic carbonates are rather unsuitable for ring-opening polymerization. Some examples have been described in the previous review, and brief outline is described below. In contrast, cyclic carbonates with larger ring (> 6) undergo ring-opening polymerization successfully by anionic, cationic, and transition metal catalysts, which has been recounted in some reviews [6-8,84-87].

In spite of the ceiling temperatures of five-membered cyclic carbonates around room temperature, only polymerizations above 100 °C have given polymers (Scheme 9) [5,6,88-97]. Regardless of the initiators, the resulting polymers are poly(ether-carbonate)s, in which contents of the ether units produced by decarboxylation usually exceed 50 mol%. Copolymerization can reduce the degree of decarboxylation [98-101]. For example, ring-opening copolymerization of EC with CL catalyzed by Sm complexes affords biodegradable poly(carbonate-ester)s without containing ether units, but the EC unit contents are around 30% [99]. EC and PC may be copolymerized with tetramethylene urea in the presence of Bu_2Mg producing polyurethanes via alternating copolymerization (Scheme 10) [102-104].

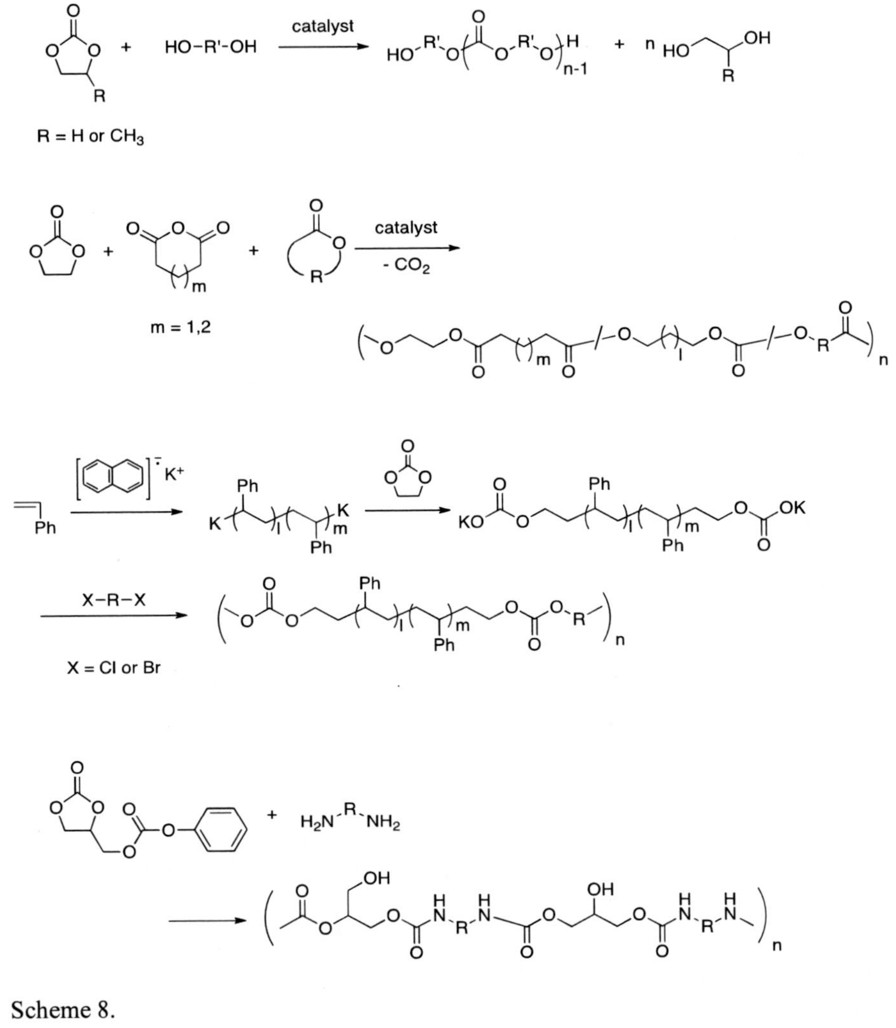

Scheme 8.

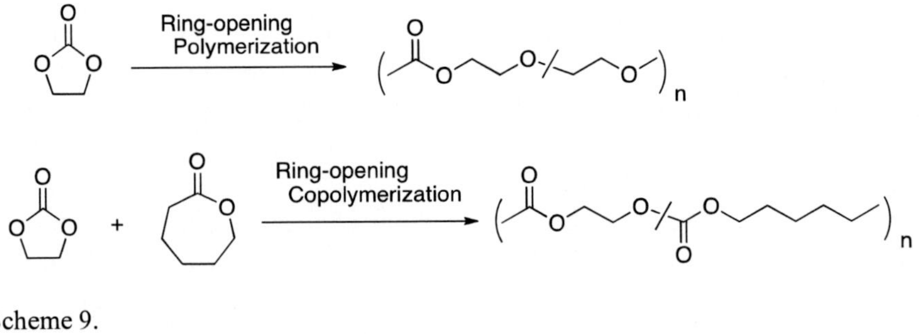

Scheme 9.

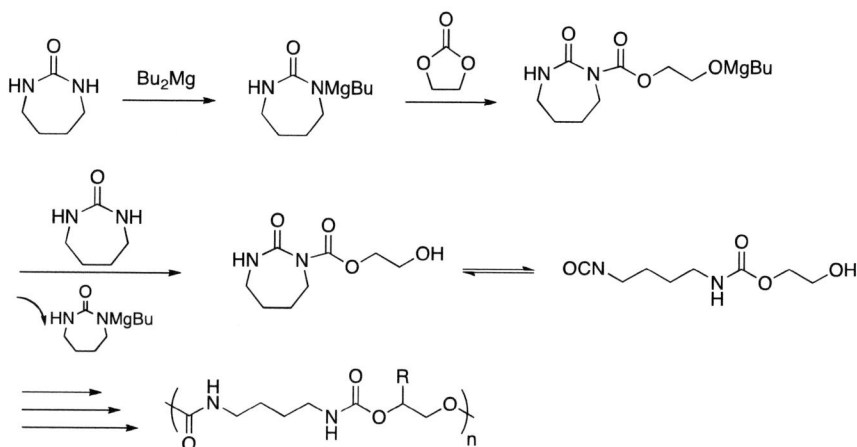

Scheme 10.

Utilizing the difference in the reactivity of five- and six-membered carbonate rings, the authors' group reported anionic polymerization of a bifunctional cyclic carbonate bearing both five- and six-membered cyclic carbonate moieties (Scheme 11) [105]. The polymerization with DBU in DMF proceeded selectively with ring-opening of the six-membered cyclic carbonate moiety to afford a polycarbonate with five-membered cyclic carbonate moieties in the side chain. Treatment of the resulting polymer by DBU under dilute conditions effectively reproduced the original monomer in a high yield.

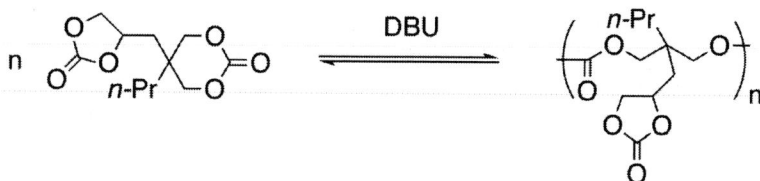

Scheme 11.

5. SYNTHESIS OF POLYMERS BEARING CARBONATE MOIETIES USING CO_2

Unique characters of cyclic carbonate moieties are; reactivity with amines, high polarity, and high refractive indexes. As a result, polymers bearing carbonate moieties in the side chain are potentially applicable as optical materials, polymeric electrolytes, adhesives, and paints. Typical synthetic methods are polymerization of monomers bearing cyclic carbonate moieties prepared from monomers bearing epoxy moieties with CO_2 and reactions of polymers bearing epoxy moieties with CO_2 (Scheme 12). Commercially available GMA is the most examined monomer for both of the methods. DOMA prepared from GMA and CO_2 has extremely high polymerizability, and as a result, its isolation and precise control over its polymerization are difficult [32-35]. DOMA is typically obtained as solutions, and solution polymerization smoothly proceed to give polyDOMA. Another approach is the conversion of epoxy moieties in polyGMA to cyclic carbonate moieties via reaction with CO_2 [106-121]. Solution phase reactions proceed effectively to transform many types of polyGMA based copolymers. This reaction may also be conducted by gas-solid phase reaction (i.e., direct incorporation of CO_2 gas into polymer film) [107,118-121]. This CO_2 fixation reaction is environmentally benign by no need of solvents or purification processes. Because catalysts must move in the film, reaction temperatures higher than T_gs are effective to attain sufficient transformation. Polymerization of epoxy groups takes place as a side reaction making the products insoluble. Crosslinking reaction may be suppressed by appropriate choice of catalysts and copolymer components [118,120,121]. Both soluble and insoluble polymers can be obtained keeping high CO_2 incorporation ratios above 94% [121]. Copolymers containing both epoxy and quaternary ammonium salt catalytic moieties were also reacted with CO_2 both by solution [108] and solid phase reactions [108,119]. The improved catalytic activity of the copolymers with self-catalytic function allows the incorporation of CO_2 into the solid-state polyGMA-based copolymer at room temperature although the incorporation efficiency was lower (30% and ca. 10% conversions by 20 days reactions under carbon dioxide and air atmospheres, respectively) [119]. The high activity was ascribed to the homogenously dispersed nature of the catalytic moieties. The advantages of the self-catalyst polymers also include no leakage of catalysts after the reactions. Improved CO_2 incorporation efficiency was attained by introducing crown-ether groups into the polymers [111].

Radical polymerization of GMA and reaction of its epoxy ring with CO_2 may be conducted concurrently [122-124], where as the aforementioned reactions are conducted in two steps. For example, reaction of GMA in the presence of AIBN and LiBr under a CO_2 atmosphere gave poly(DOMA-co-GMA), whose unit ratios may be controlled by reaction conditions [122]. When 1,4-dioxane was used as the solvent, the resulting polymer precipitated as the progress of the reaction, and could be collected by filtration and washing [123]. Spherical particles may be obtained in the presence of cellulose acetate as a dispersing agent [124]. The epoxy moieties in poly(DOMA-*co*-GMA) may be selectively hydrolyzed under acidic conditions, and the remained carbonate moieties may be subsequently transformed to hydroxyurethane moieties via reactions with amines [122,124].

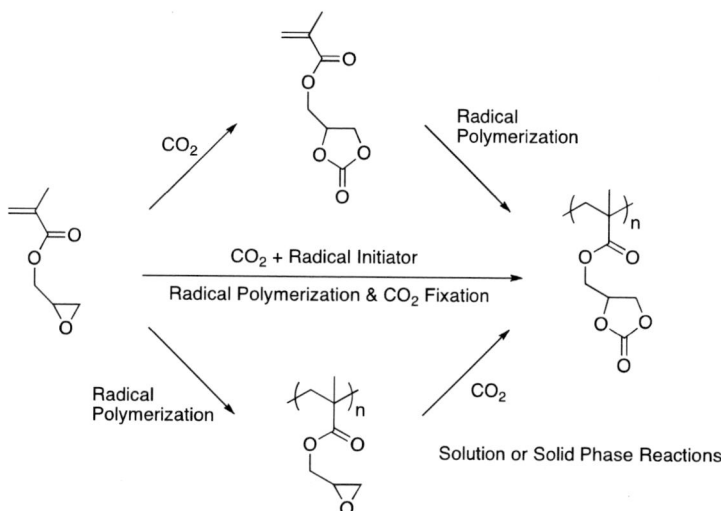

Scheme 12.

Park et al. reported miscibilities of polyDOMA-based copolymers [112-117]. These copolymers are miscible with various polymers such as poly(alkyl methacrylate)s [112,113], poly(vinyl chloride) [112], and poly(styrene-*co*-acrylonitrile) [115], whereas analogous polyGMA-based copolymers are immiscible. Based on the miscibility improving transparency of polymers, polymers containing cyclic carbonate and cinnamic ester groups are photo-crosslinkable, and has good photosensitivity [116,117]. Other potential applications of carbonate pendant polymers include polymer electrolyte based on the high dielectric constant of cyclic carbonate structure, optical materials, adhesives, and solid support for enzyme immobilization. Application of polymers bearing cyclic carbonate moieties as polymer electrolyte for lithium-ion batteries was examined by Golden et al. [33] and Wegner et al. [125]. PolyDOMA is too hard for polymer electrolyte, but polyDOA an acrylate analogue and polymers bearing alkylene spacers are more suitable. The ionic conductivity was measured for elastomeric gels of polymers and PC for plasticization containing lithium salts to be in the order of $10^{-3}-10^{-5}$ S/cm.

Evidence for Li$^+$-cyclic carbonate interaction was investigated by ^{13}C CP-MAS NMR spectroscopic analysis, for which poly(vinylene carbonate) was employed as a model polymer. The interaction of more than one carbonyl group per Li$^+$ was suggested by integral ratio of the carbonyl peaks attributable to ligated and unligated carbonyls. Perspective of carbonate polymers for other applications was represented in an earlier review [32].

CONCLUSION

This chapter described synthesis and application of polymers obtained based on the reaction of epoxides and CO_2. Polyhydroxyurethanes are unique polymers and may contain various functional groups. Potential applications involve paints, packaging materials, biomaterials, and electronic devices. Polymers bearing cyclic carbonate groups may be prepared by various procedures. Their applications to polymer electrolytes for lithium-ion batteries have been examined, and various potential applications are also investigated. Both of the carbonate-based polymers are obtained by simple methods under mild conditions. Continued developments in this area have the potential for the large-scale production of these polymers. Transformation of carbon dioxide into useful materials is an important subject in view of the demand on new petroleum resources. Industrial fabrications of aromatic and aliphatic polycarbonates using CO_2 were recently started [1], and further efforts will open the door for fabrication of various commercial polymers using CO_2 via green processes.

REFERENCES

[1]   Fukuoka, S. (2012). *Non-phosgene polycarbonate from CO₂ - industrialization of green chemical process-;* Nova Science Publishers: Hauppauge, NY.

[2]   Inoue, S. (1976). *Chemtech*, 588-594.

[3]   Rokicki, A.&Kuran, W. *J. Macromol. Sci. Rev. Macromol. Chem.* 1981, *C21*, 135-186.

[4]   Tsuda, T. (1995). *Gazz. Chim. Ital., 125*, 101-110.

[5]   Super, M. S. & Beckman, E. J. (1997). *Trends Polym. Sci., 5*, 236-240.

[6]   Kuran, W. (1998). *Progr. Polym. Sci., 23*, 919-992.

[7]   Rokicki, G. (2000). *Progr. Polym. Sci.*, *25*, 259-342.
[8]   Darensbourg, D. J. & Holtcamp, M. W. (1996). *Coord. Chem. Rev.*, *153*, 155-174.
[9]   Beckman, E. J. (2004). *J. Supercrit. Fluid*, *28*, 121-191.
[10] Ochiai, B.& Endo, T. (2005). *Progr. Polym. Sci.*, *30*, 183-215.
[11] Aresta, M.& Dibenedetto, A. (2007). *Dalton Trans.*, 2975-2992.
[12] Sakakura, T., Choi, J.-C. & Yasuda, H. (2007). *Chem. Rev.*, *107*, 2365-2387.
[13] Kihara, N., Hara, N.& Endo, T. (1993). *J. Org. Chem.*, *58*, 6198-6202.
[14] Barkakaty, B.,Morino, K.,Sudo, A.& Endo, T. (2010). *Green Chem.*, *12*, 42-44.
[15] Barkakaty, B., Morino, K., Sudo, A. & Endo, T. (2011). *J. Polym. Sci., Part A: Polym. Chem.*, *49*, 545-549.
[16] Aoyagi, N., Furusho, Y. & Endo, T. (2012). *Chem. Lett.*, *41*, 240-241.
[17] Aoyagi, N., Furusho, Y. & Endo, T. (2013). *J. Polym. Sci., Part A: Polym. Chem.*, *51*, 1230-1242.
[18] Nishikubo, T., Kameyama, A., Yamashita, J.,Tomoi, M. & Fukuda, W. (1993). *J. Polym. Sci., Part A: Polym. Chem.*, *31*, 939-947.
[19] Nishikubo, T., Kameyama, A., Yamashita, J., Fukumitsu, T., Maejima, C. & Tomoi, M. (1995). *J. Polym. Sci., Part A: Polym. Chem.*, *33*, 1011-1017.
[20] Du, Y.,Cai, F., Kong, D.-L. & He, L. N. (2005). *Green Chem.*, *7*, 518-523.
[21] Ochiai, B. & Endo, T. (2007). *J. Polym. Sci., Part A: Polym. Chem.*, *45*, 5673-5678.
[22] Trost, B. M. & Angle, S. R. (1985). *J. Am. Chem. Soc.*, *107*, 6123-6124.
[23] Ochiai, B., Matsuki, M., Nagai, D., Miyagawa, T. & Endo, T. (2005). *J. Polym. Sci., Part A: Polym. Chem.*, *43*, 584-592.
[24] Cho, I. & Lee, T. W. (1989). *Makromol. Chem. Rapid Commun.*, *10*, 453-456.
[25] Joumier, J. M., Fournier, J., Bruneau, C. & Dixneuf, P. H. (1991). *J. Chem. Soc. Perkin Trans.*, 3271-3274.
[26] Uemura, K., Kawaguchi, T., Takayama, H., Nakamura, A., & Inoue, Y. (1999). *J. Mol. Cat. A Chem.*, *139*, 1-9.
[27] Ochiai, B., Sano, Y.& Endo, T. (2005). *J. Network Polym. Jpn.*, *26*, 132-137.
[28] Pritchard, W. W. (1950). *US:2,511,942.*
[29] Asahara, T., Seno, M. & Imai, T. (1973). *Seisan Kenkyu*, *25*, 297-299.
[30] Webster, D. C. & Crain, A. L. (2000). *Progr. Org. Coat.*, *40*, 275-282.

[31] D'Allelio, G. F. & Huemmer, T. (1967). *J. Polym. Sci. Part A-1*, *5*, 307-321.

[32] Brosse, J. C. & Couvret, D. *Makromol.* (1990). *Chem. Rapid Commun.*, *11*, 123-128.

[33] Golden, J. H., Chew, B. G. M.,Zax, D. B., DiSalvo, F. J., Fréchet, J. M. J. & Tarascon, J.-M. (1995). *Macromolecules*, *28*, 3468-3570.

[34] Kihara, N.& Endo, T. (1992). *Makromol. Chem.*, *193*, 1481-1492.

[35] Berchtold, K. A. ,Nie, J.,Stansbury, J. W. & Bowman, C. N. (2008). *Macromolecules*, *41*, 9035-9043.

[36] Baba, A. ,Kashiwagi, H. & Matsuda, H. (1985). *Tetrahedron Lett.*,*26*, 1323-1324.

[37] Darensbourg, D. J., Moncada, A. I. & Wei, S.-H. (2011). *Macromolecules*, *44*, 2568-2576.

[38] Parrish, J. P., Salvatore, R. N. & Jung, K. W. (2000). *Tetrahedron*, *56*, 8207-8237.

[39] Cardillo, G., Orena, M., Porzi, G. & Sandri, S. (1981). *J. Chem. Soc. Chem. Commun.*, 465-466.

[40] Bongini, A., Cardillo, G., Orena, M., Porzi, G. & Sandri, S. (1982). *J. Org. Chem.*, *47*, 4626-4633.

[41] Baizer, M. M., Clark, J. R. & Smith, E. (1957). *J. Org. Chem.*, *22*, 1706.

[42] Stout, E. I., Doane, W. M., Shasha, B. S., Russell, C. R. & Rist, C. E. (1967). *Tetrahedron Lett.*, *8*, 4481.

[43] Sklavounos, C., Goldman, I. M. & Kuhla, D. E. (1980). *J. Org. Chem.*, *45*, 4239-4240.

[44] Tomita, H., Sanda, F. & Endo, T. (2001). *J. Polym. Sci., Part A: Polym. Chem.*, *39*, 3678-3685.

[45] Tomita, H., Sanda, F. & Endo T. (2001). *J. Polym. Sci., Part A: Polym. Chem.*, *39*, 851-859.

[46] Tomita, H., Sanda, F. & Endo T. (2001).*J. Polym. Sci., Part A: Polym. Chem.*, *39*, 162-168.

[47] Tomita, H., Sanda, F. & Endo T. (2001).*J. Polym. Sci., Part A: Polym. Chem.*, *39*, 4091-4100.

[48] Whelan Jr., J. M., Hill, M. & Cotter, R. J. *US:3,072, 613*, 1963.

[49] Mikheev, V. V.,Svetlakov, N. V.,Sysoev, V. A. & Brus'ko, N. V. (1983). *Deposited Doc.* 1982, *SPSTL 41 Khp-D82*, 8. *Chem. Abstr.*, *98*, 127745a.

[50] Rokicki, G. & Lewandowski, M. (1987). *Angew. Makromol. Chem.*, *148*, 53-66.

[51] Rokicki, G. & Lazinski, R. (1989). *Angew. Makromol. Chem.*, *170*, 211-225.

[52] Rokicki, G. & Czajkowska, J. (1989). *Polimery*, *34*, 140-147. *Chem. Abstr.*1991, *114*, 186119.

[53] Rokicki, G. & Wojciechowski, C. (1990). *J. Appl. Polym. Sci.*, *41*, 647-659.

[54] Kihara, N. & Endo T. (1993). *J. Polym. Sci., Part A: Polym. Chem.*, *31*, 2765-2773.

[55] Kihara, N., Kushida, Y. & Endo, T. (1996). *J. Polym. Sci., Part A: Polym. Chem.*, *34*, 2173-2179.

[56] Steblyanko, A., Choi, W.-M. Sanda, F. & Endo, T. (2000). *J. Polym. Sci., Part A: Polym. Chem.*, *38*, 2375-2380.

[57] Tomita, H., Sanda, F. & Endo, T. (2001). *J. Polym. Sci., Part A: Polym. Chem.*, *39*, 860-867.

[58] Kim, M.-R., Kim, H.-S., Ha, C.-S., Park, D.-W. & Lee, J.-K. (2001). *J. Appl. Polym. Sci.*, *81*, 2735-2743.

[59] Al-Azemi, T. F. & Bisht, K. S. (2002). *Polymer*, *43*, 2161-2167.

[60] Tamami, B., Sohn, S. & Wilkes, G. L. (2004). *J. Appl. Polym. Sci.*, *92*, 883-891.

[61] Suzuki, A., Nagai, D., Ochiai, B. & Endo, T. (2004). *J. Polym. Sci., Part A: Polym. Chem.*, *42*, 5983-5989.

[62] Ochiai, B., Satoh, Y. & Endo, T. (2005). *Green Chem.*, *7*, 765-767.

[63] Ochiai, B., Nakayama, J., Mashiko, M., Kaneko, Y. & Endo, T. (2005). *J. Polym. Sci., Part A: Polym. Chem.*, *43*, 5899-5905.

[64] Ochiai, B., Inoue, S. & Endo, T. (2005). *J. Polym. Sci., Part A: Polym. Chem.*, *43*, 6613-6618.

[65] Ochiai, B., Sato, S. & Endo, T. (2007). *J. Polym. Sci., Part A: Polym. Chem.*, *45*, 3400-3407.

[66] Ochiai, B., Sato, S. & Endo, T. (2007). *J. Polym. Sci., Part A: Polym. Chem.*, *45*, 3408-3414.

[67] Ochiai, B., Satoh, Y. & Endo, T. (2009). *J. Polym. Sci., Part A: Polym. Chem.*, *47*, 4629-4635.

[68] Ochiai, B., Koda, K. & Endo, T. (2012). *J. Polym. Sci., Part A: Polym. Chem.*, *50*, 47-51.

[69] Garipov, R. M., Sysoev, V. A., Mikheev, V. V., Zagidullin, A. I., Deberdeev, R. Ya., Irzhak, V. I. & Berlin, Al. Al. (2003). *Dokl. Phys. Chem.*, *393*, 61-64.

[70] Bähr, M.,Bitto, A. & Mülhaupt, R. (2012). *Green Chem.*, *14*, 1447-1454.

[71] Kathalewar, M. S., Joshi, P. B.,Sabnis, A. S. & Malshe, V. C. (2013). *RSC Adv.*, *3*, 4110-4129.

[72] Kihara, N., Makabe, K.& Endo, T. (1996). *J. Polym. Sci., Part A: Polym. Chem.*, *34*, 1819-1822.

[73] El-Giamal, M. F. & Schulz, R. C. (1976). *Makromol. Chem.*, *177*, 2259-2269.

[74] McCabe, R. W. & Taylor, A. (2002). *Chem. Commun.*, 934-935.

[75] Rokicki, G. & Piotrowska, A. (2002). *Polymer*, *43*, 2927-2935.

[76] Ochiai, B.& Utsuno, T. (2013). *J. Polym. Sci., Part A: Polym. Chem.*, *51*, 523-533.

[77] Ubaghs, L., Fricke, N., Keul, H. & Höcker, H. (2004). *Macromol. Rapid Commun.*, *25*, 517-521.

[78] Rokicki, G. & Kowalczyk, T. (2000). *Polymer*, *41*, 9013-9031.

[79] Pawlowski, P. & Rokicki, G. (2004). *Polymer*, *45*, 3125-3137.

[80] Schwenk, V. E., Gulbins, K., Roth, M.,Benzing, G. Maysenhölder, R. & Hamann, K. (1962). *Makromol. Chem.*, *51*, 53-69.

[81] Kricheldorf, H. R. & Petermann, O. (2002). *J. Polym. Sci., Part A: Polym. Chem.*, *40*, 4356-4367.

[82] Rokicki, G. & Jezewski, P. (1988). *Polym. J.*, *20*, 499-509.

[83] Rokicki, G., Pawlicki, J. & Kuran, W. (1985). *Polym. J.*, *17*, 509-516.

[84] Endo, T. & Sanda, F. (1996). *Ring-opening polymerization, anionic (with expansion in volume).* In: *Polymeric Materials Encyclopedia.*Salamone, J. C., Ed.CRC Press Inc., Boca Raton, 7550-7554.

[85] Endo, T. & Sanda, F. (1996). *Ring-opening polymerization, cationic (with expansion in volume).* In: *Polymeric Materials Encyclopedia.*Salamone, J. C., Ed.CRC Press Inc., Boca Raton, 7554-7560.

[86] Keul, H. & Höcker, H. (2000). *Macromol. Rapid Commun.*,*21*, 869-883.

[87] Okada, M. (2002). *Progr. Polym. Sci.*, *27*, 87-133.

[88] Soga, K., Tazuke, Y., Hosoda, S.& Ikeda, S. (1977). *J. Polym. Sci., Polym. Lett. Ed.*, *15*, 219-229.

[89] Sakai, S., Suzuki, M., Aono, T., Hasebe, K., Kanbe, H., Hiroe, M., Kakei, T., Fujinami, T. & Takemura, H. (1984). *KobunshiRonbunsyu*, *41*, 151-158.

[90] Vogdanis, L. & Heitz, W. (1986). *MakromolChem Rapid Commun*, *7*, 543-547.

[91] Harris, R. F. (1989). *J. Appl. Polym. Sci.*, *38*, 463-476.

[92] Harris, R. F. & McDonald, L. A. (1989). *J. Appl. Polym. Sci.*, *37*, 1491-1511.
[93] Vogdanis, L., Martens, B., Uchtmann, H., Hensel, F. & Heitz, W. (1990). *Makromol. Chem.*, *191*, 465-472.
[94] Storey, R. F. & Hoffman, D. C. (1992). *Macromolecules*, *25*, 5369-5382.
[95] Soós, L., Deák, G. Y., Kéki, S. & Zsuga, M. (1999). *J. Polym. Sci.*, *Part A: Polym. Chem.*, *37*, 545-550.
[96] Lee, J. C. & Litt, M. H. (2000). *Macromolecules*, *33*, 1618-1627.
[97] Kadokawa, J., Iwasaki, Y. & Tagaya H. (2002). *Macromol. Rapid Commun.*, *23*, 757-760.
[98] Kuran, W. & Listos, T. (1992). *Makromol. Chem.*, *193*, 945-956.
[99] Evans, W. J. & Katsumata, H. (1994). *Macromolecules*, 27, 4011-4013.
[100] Agarwal, S., Naumann, N. & Xie, X.-L. (2002). *Macromolecules*, *35*, 7713-7717.
[101] Shirahama, H., Kanetani, A. & Yasuda, H. (2000). *Polym. J.*, *32*, 280-286.
[102] Schmitz, F., Keul, H. & Höcker H. (1997). *Macromol. Rapid Commun.*, *18*, 699-706.
[103] Schmitz, F., Keul, H. & Höcker, H. (1998). *Polymer*, *39*, 3179-3186.
[104] Ubaghs, L., Novi, C., Keul, H. & Höcker H. (2004). *Macromol. Chem. Phys.*, *205*, 888-896.

[105] Endo, T., Kakimoto, K., Ochiai, B. & Nagai, D. (2005). *Macromolecules*, *38*, 8177-8182
[106] Kihara, N. & Endo, T. (1992). *Macromolecules*, *25*, 4824-4825.
[107] Kihara, N. & Endo, T. (1994). *J. Chem. Soc. Chem. Commun.*, 937-938.
[108] Kihara, N. & Endo, T. (1994). *Macromolecules*, *27*, 6239-6244.
[109] Sakai, T., Kihara, N. & Endo, T. (1995). *Macromolecules*, *28*, 4701-4706.
[110] Yamamoto, S., Hayashi, T., Kawabata, K., Moriya, O. & Endo, T. (2002). *Chem. Lett.*, 816-817.
[111] Yamamoto, S., Moriya, O. & Endo T. (2003). *Macromolecules*, *36*, 1514-1521.
[112] Park, S.-Y., Park, H.-Y., Lee, H.-S., Park, S.-W., Ha, C.-S. & Park, D.-W. (2001). *J. Polym. Sci., Part A: Polym. Chem.*, *39*, 1472-1480.
[113] Park, S.-Y., Lee, H.-S., Ha, C.-S. & Park, D.-W. (2001). *J .Appl. Polym. Sci.*, *81*, 2161-2169.
[114] Park, S.-Y., Park, H.-Y., Park, D.-W. & Ha, C.-S. (2002). *J. Macromol. Sci., Pure Appl. Chem.*, *39*, 573-589.

[115] Park, S.-Y., Park, H.-Y., Woo, H.-S., Ha, C.-S. & Park, D.-W. (2002). *Polym. Adv. Technol.*, *13*, 513-521.

[116] Park, S.-Y., Park, H.-Y., Lee, H.-S., Park, S.-W. & Park, D.-W. (2003). *Opt. Mater.*, *21*, 331-335.

[117] Lee, D.-W., Hur, J.-H., Kim, B.-K., Park, S.-W. & Park, D.-W. (2003). *J. Ind. Eng. Chem*, *9*, 513-517.

[118] Ochiai, B., Iwamoto, T., Miyagawa, T., Nagai, D. & Endo, T. (2004). *J. Polym. Sci., Part A: Polym. Chem.*, *42*, 3812-3817.

[119] Ochiai, B., Iwamoto, T., Miyagawa, T., Nagai, D. & Endo, T. (2004). *J. Polym. Sci., Part A: Polym. Chem.*(2004, *42*, 4941-4947.

[120] Ochiai, B., Iwamoto, T., Miyazaki, K. & Endo T. (2005). *Macromolecules*, *38*, 9939-9943.

[121] Ochiai, B., Iwamoto, T. & Endo, T. (2006). *Green Chem.*, *8*, 138-140.

[122] Ochiai, B., Hatano, Y. & Endo, T. (2008). *Macromolecules*, *41*, 9937-9939.

[123] Ochiai, B., Hatano, Y. & Endo, T. (2009). *J. Polym. Sci., Part A: Polym. Chem.*, *47*, 3170-3176.

[124] Ochiai, B. & Nakayama, T. (2010). *J. Polym. Sci., Part A: Polym. Chem.*, *48*, 5382-5390.

[125] Britz, J., Meyer, W. H. & Wegner, G. (2007). *Macromolecules*, *40*, 7558-7565

In: Carbonates   ISBN: 978-1-62948-178-4
Editors: B.A. Hughes, T.C. Wagner   © 2013 Nova Science Publishers, Inc.

Chapter 3

PHYSICAL REDISTRIBUTION OF CALCIUM CARBONATE IN A SOIL PORE SYSTEM: AN EXPERIMENT USING SOIL MICROMORPHOLOGY AND IMAGE ANALYSIS

*Laura Gargiulo[*1], Giacomo Mele[1] and Fabio Terribile[2†]*

[1]Istituto per i Sistemi Agricoli e Forestali del Mediterraneo,
Consiglio Nazionale delle Ricerche, Ercolano, Italy
[2]Dipartimento di Agraria, Università di Napoli Federico II,
Portici, Italy

ABSTRACT

Calcium carbonate sedimentation plays an important role in the soil structure development. Soil structure is a key factor for the soil functions, such as sustaining plant productivity, regulating and partitioning water and solute flow, and maintaining civil engineering works. In order to improve soil structure the calcium carbonate is added to the soil by means liming procedure in order to improve soil chemical and physical properties, neutralizing soil acidity and increasing activity of soil bacteria, and improving the soil structure. A large bulk of scientific

[*] E-mail: lauragargiulo@alice.it.
[†] E-mail: fabio.terribile@unina.it.

literature addresses the relations between the forms of carbonate redistribution in the development of calcic and petrocalcic horizons and the water regime of calcareous hydromorphic soils, but little is still known concerning the underlying physical mechanisms of the effect of carbonate sedimentation on soil pore system. In this work we attempted to investigate physical mechanisms of soil pore development as consequence of the addition of calcium carbonate ($CaCO_3$) on two soils with different shrinkage-swelling capacity subjected to several wetting and drying cycles. Analysis was conducted using 2D soil image analysis and soil micromorphology. Our results showed changes in the pore size distribution, in some cases very large, and allowed the identification of specific mechanisms of pore modification induced by micrite pedofeatures produced by the mobilization in suspension of $CaCO_3$. These physical mechanisms were triggered by $CaCO_3$ segregations, which induced a pore size redistribution fragmenting the pore space, and $CaCO_3$ coatings, which seemed to induce a cumulative effect on porosity cementing the walls of newly-formed pores in the soil samples with high shrinkage-swelling capacity.

Our results, even if obtained on experimental samples, give a contribution in the understanding of the physical role of $CaCO_3$ pedofeatures in pore formation in soils in field and show the need to reassess physical simulation tests in order to quantitatively investigate combined effects of factors influencing soil structure formation.

Keywords: Soil structure; soil image analysis; soil pore formation; agents of soil pore development; wetting-drying cycles; soil micromorphology; $CaCO_3$ pedofeatures

ABBREVIATIONS

Calcium carbonate = $CaCO_3$
Wetting – drying cycles = W/D cycles
Pore size distribution = PoSD
Low shrink-swell soil = LS soil
High shrink-swell soil = HS soil

INTRODUCTION

Soil structure is the complex system of soil pore space resulting from the spatial organization of the solid phase (Lal, 1991) and is a critical physical property that affects soil ability in maintaining agricultural productivity (Hillel, 1980) as well as local and global environmental quality (Bronick and Lal, 2005).

The formation of soil structure is the result of the actions and interactions of numerous physical, chemical and biological factors with intricate feedback mechanisms (Six et al., 2004) making difficult to understand the specific effects of each single factor.

Calcium carbonate is, among all inorganic agents influencing soil structure, especially important in soils due to its widespread natural occurrence in many soil diagnostic horizons. For example, $CaCO_3$ coatings cement together sand and silt particles during the formation of both calcic and petrocalcic horizons (Gile et al., 1966; Lal, 2006).

Carbonate redistributes in soil because of a variety of processes: biological, producing the formation of calcitized cells, (e.g. Jaillard et al., 1991), chemical (e.g. Delmas et al., 1987), with surface dissolution of carbonate minerals, or purely physical, as a result of wetting/drying (W/D) cycles in arid or Mediterranean environments (e.g. Kaemmerer et al., 1991; Kaemmerer & Revel, 1996).

Baghernejad and Dalrymple (1993) showed that $CaCO_3$ can be mobilized in the pore networks as suspension and argued that such physical mechanism plays an important role in the process of structure formation of calcic horizons, but real pore measurements were not done.

Actually direct investigations of the soil pore system, such as soil image analysis, are available and provide valid tools to analyze both shape and size distribution of pores (Mele et al., 1999; Velde, 1999; Pagliai and Vignozzi, 2002). Unfortunately soil image analysis is still little used to investigate the role of the numerous factors influencing pore architecture. More recently, Falsone et al. (2010) quantified the effect of calcification on the soil pore system, indirectly, analysing this latter at a nanometer scale by means of nitrogen adsorption and mercury porosimetry. Moreover, almost all micromorphological studies on relationship between $CaCO_3$ pedofeatures and soil pore system (Khormali et al., 2006; Shankar and Achyuthan, 2007) are based on undisturbed natural soil samples and, although can provide useful information on soil genesis, do not allow to discriminate the specific mechanisms by which those pedofeatures can induce the pore development.

In such framework, given the importance of physical redistribution of calcium carbonate for formation and preservation of soil structure, we chose to perform a laboratory experimental work of wetting/drying cycles on soil samples prepared adding $CaCO_3$.

The aim was to identify and quantify, coupling soil micromorphology and pore image analysis, the physical effects of $CaCO_3$ on pore development in two soils with very different shrinking-swelling capacity, as the degree of soil shrinkage and swelling largely determines the potential for soil structure development (Loveday, 1972).

In order to better identify only the effects of the added substances combined with the W/D cycles, have been chosen two soils characterized by low content of other structuring agents such as organic matter and biota, and not evident self-structuring (e.g. self-mulching) behavior typical of soils having high content of expandable clay minerals (e.g. Vertisols).

MATERIALS AND EXPERIMENTAL DESIGN

The experiment was performed on two soils. The soil with low shrinkage-swelling capacity (LS) was a *Psamment* developed over a recent sand dune at Palinuro (Salerno, South Italy, 40°02'18"N - 15°17'33"E). The soil with high shrinkage-swelling (HS) capacity was an *Orthent* developed over fine sediments sampled on an alluvial terrace near the dam of the Alento River (Salerno, South Italy, 40° 19'34"N - 15 ° 06'08"E).

Fine-grain $CaCO_3$ (<10 μm) powder (supplied by Sigma-Aldrich Inc.) was used in order to obtain a near-colloidal suspension of micrite crystals resembling natural carbonate in soils such as those having illuvial calcic horizons (Lal, 2006) or those developing over young (carbonate rich) alluvial deposits. In these soils carbonate redistribution (generally through coatings) is an important process and it can occur by both evaporation and mechanical migration processes (Durand et al., 2010).

$CaCO_3$ was dry-mixed at three different concentrations (0.5g/kg, 5.0g/kg, 50g/kg) with both soils. The concentrations were selected based on a review of related literature (Muneer and Oades, 1989). Three replicate polypropylene pots (6.0 cm high, 6.5 cm in diameter) for each study case and three untreated control pots for each soil were prepared. In order to trigger soil structure development the samples were put in a tank and subjected to nine wetting/drying (W/D) cycles, consisting of a wetting phase of 24 hours at 25 °C and drying phase of 4 days at 40°C. Sample dry weight returned to the initial value after the 4 days drying period. During W/D cycles moisture ratio ranged, on average, between 0.40 and 1.50 for the HS soil and between 0.05 and 0.55 for the LS soils. In order to avoid possible soil structure artefacts induced by drop impact or runoff, wetting (with deionised water) was performed via capillary action from the bottom of the pots. Nine cycles were used based on previous research that indicated stabilization of the pore size distribution after four cycles in clay loam and sandy loam soils (Gargiulo, 2008).

METHODS

Soil Characterization

Prior to the analysis and experiment, the two soils were dried at 40 °C for 72 hours and sieved to 2 mm. Soils were analysed for texture by sieving a humid sample for the fractions between 0.2 and 2 mm, and by sedimentation (pipette method) for <0.2 mm fractions (Gee and Or, 2002).

Following standard methods completed soil chemical analyses. Soil pH was determined potentiometrically with a pHmeter (10pH/ISE, Beckman) in soil-water suspensions (ratio of 1:2.5) (Peech, 1965). Organic carbon content was determined with the Walkley and Black (1934) method, by means of organic matter oxidation with potassium bichromate, in the presence of sulphuric acid. Electrical conductivity was measured in soil-water suspensions (ratio of 1:5) using a conductivity meter (microCM 2201, CRISON) (Rhoades, 1996). Total carbonates were determined using a Dietrich-Fruehling calcimeter (Loeppert and Suarez, 1996); cation exchange capacity (CEC) was determined with $BaCl_2$ (Summer and Miller, 1996).

Mineralogical analysis was performed by X-ray diffraction (XRD) (Wilson, 1987). Samples were dispersed and separated into different grain size classes through sieving to obtain sand (50 μm - 2 mm), and centrifuged to obtain clay (<2 μm). The clay was saturated with $CaCl_2$ and washed with water and acetone to remove chloride. Spectra were determined using a Rigaku Geigerflex D / Max IIIC diffractometer with CuKa radiation and a Ni filter, at 40 kW and 25 mA. Oriented clay samples solvated in ethylene glycol were analyzed to identify expandable secondary minerals. Powder samples from the sand fraction were analyzed randomly in order to define the primary mineral components.

Soil plasticity was also measured on thoroughly puddled samples of the two soils at a water content where maximum plasticity is expressed, according to the field method described in the Soil survey manual (Soil Survey Division Staff, 1993).

The shrink-swell potential of the soils was evaluated by the coefficient of linear extensibility on a whole-soil base ($COLE_{ws}$) (USDA, 2004),

$$COLE_{ws} = (Db_{d<2mm}/Db_{33<2mm})^{1/3} - 1 \tag{1}$$

where $Db_{d<2mm}$ = Bulk Density, oven-dry or air-dry, on a <2-mm base (g cm^{-3}), $Db_{33<2mm}$ = Bulk Density at 33-kPa water content on a <2-mm base (g cm^{-3}).

The shrinkage dynamics of the two soils were investigated too, using the method of Brasher (1966) to obtain the shrinkage characteristic curves (Groenevelt and Grant, 2004).

Two-Dimensional Image Analysis

A mixture of acetone and polyester resin (Crystic 17449, Scott-Bader Ltd.) was added with fluorescent dye (Uvitex OB, Ciba Ltd.), having a spectral emission in the blue band under UV illumination (365 nm). The three replicates from each treatment were saturated with the acetone-resin mixture under a moderate vacuum. This procedure yielded a low viscosity mixture for optimal resin penetration into the pore networks (Fitzpatrick, 1993). After resin polymerization the soil blocks were cut into regular parallelepipeds. Digital images (10 μm pixel resolution) of the four vertical sections (3 × 5 cm) were acquired under UV illumination. A Nikon D200 camera was used, controlled by a PC using Nikon Capture 4.1 software. To merge the variability of the three replicates in each treatment, the four images of the vertical sections of each sample (Figure 1) were placed side by side to obtain a single large 2D image (36 × 5 cm) consisting of all twelve vertical sections of the treatment.

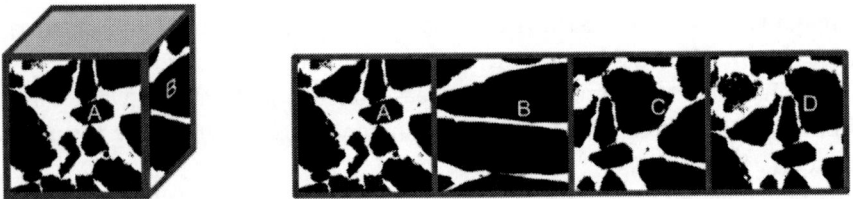

Figure 1. Scheme of analysed images of a soil block. Vertical sections of a soil block (left) and the correspondent binary image, containing the 4 vertical sections A, B, C and D that was analyzed (right).

Images were pre-processed and segmented using a technique of supervised "thresholding" using Corel Photo-Paint X3, in order to obtain binary images where the two separate solid and pore phases are in black and white, respectively.

Image analysis was performed using Solicon - PC Version 1.0 software (Cattle et al., 2000) to determine porosity values. Pore size distribution was determined by image analysis using Micromorph 1.4 (TRANSVALOR 2000), through the successive application of the "opening" algorithm (Horgan, 1998; Serra, 1982), which classifies the pore phase according to the spacing from the walls.

Micromorphological Analysis

After acquisition of digital images, each soil block was used to prepare one thin section (Fitzpatrick, 1993) which was analysed by optical microscopy using transmitted light (TL), cross polarized light (XPL) and incident light (IL) to identify the different micromorphological features. Micromorphological analysis was performed to detect features relevant for understanding the influence of $CaCO_3$ on pore development. Such features were described following FitzPatrick (1993).

The proportion of all micromorphological features of interest was estimated by the point counting technique (McKeague et al., 1980). A minimum of 3000 points (1000 per thin section) was counted for each feature, in order to obtain the percentage of the solid phase area represented by that feature. Standard error e was calculated according to the following formula (Murphy, 1983):

$$e = [((N\text{-}F)/N)*(F/N)*(1/(N\text{-}1))]^{1/2} \qquad (2)$$

where N is the total number of points counted (3000) and F is the total number of points counted as a micromorphological feature.

RESULTS

Results presented here focus on the highest concentration tested. However, results obtained at lower concentrations were in agreement with trends observed at the highest one.

Soil Characterization

The main properties of the two soils are given in Table 1.

Table 1. Properties of the two soils

Soils	Texture			pH	EC[a] (µS cm^{-1})	OC[b] (g Kg^{-1})	Total carbonate (g Kg^{-1})	CEC[c] (cmol(+) Kg^{-1})
	Sand (%)	Silt (%)	Clay (%)					
Low shrink-swell	95	3	2	8.7	71	1.3	79.4	1.3
High shrink-swell	21	59	20	7.7	382	7.4	102.3	9.8

[a]electrical conductivity.
[b]organic carbon.
[c]cation exchange capacity.

The texture of the LS soil and HS soil were, respectively, sandy and silty loam.

LS soil was strongly alkaline while the HS soil was slightly alkaline and both were nonsaline. The CEC of the LS soil was very low (1.3 cmol(+) Kg^{-1}), in agreement with its sandy texture, while the CEC of the HS soil was moderate (9.8 cmol(+) Kg^{-1}). Exchangeable sodium percentage (ESP) was 12% for the LS soil, which is clearly consistent with the environmental setting of this site (recent sand dune), and only about 2% for the HS soil, resulting both non-sodic soils (USDA, 1954). The organic carbon concentration was low in both soils, although higher in the HS soil. The carbonate levels in the LS soil and HS soil were much higher (79.4 g kg^{-1} and 102.3 g kg^{-1}, respectively), corresponding to classifications of moderately calcareous and calcareous, respectively. Quartz (peaks at 0.43 nm and 0.33 nm), calcite (peak at 0.30 nm) and albite (peak at 0.32 nm) were identified in the LS soil. Three clay minerals were identified in the HS soil: interstratified kaolinite – smectite (peaks at 0.83 nm and 1.78 nm after ethylene glycol treatment), kaolinite (peaks at 0.71 nm and 0.38 nm corresponding to d(001) and d(002) reflections) and illite (peaks at 0.99 nm, 0.51 nm and 0.33 nm corresponding to d(001), d(002) and d(003) reflections).

The LS soil was not-plastic, the HS soil slightly plastic. The coefficient of linear extensibility (COLE$_{std}$) value resulted 0.060 for the HS soil and 0.004 for the LS soil, corresponding to classifications of high shrink-swell potential and low shrink-swell potential, respectively, according to National Soil Survey Handbook (USDA, 2010).

In figure 2 the shrinkage characteristic curves of the two soils are shown.

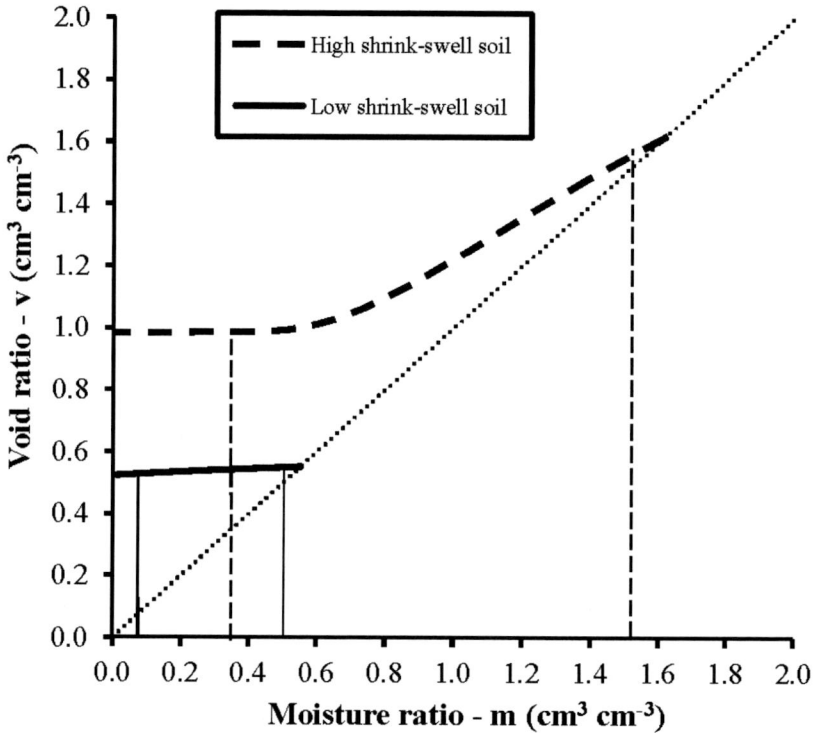

Figure 2. Shrinkage curves of low shrink-swell soil and high shrink-swell soil. Vertical solid (for low shrink-swell soil) and dashed (for high shrink-swell soil) lines indicate minimum and maximum moisture ratio values achieved during the W/D cycles. The dotted line is the reference line at saturation.

Minimum and maximum moisture ratio values reached during W/D cycles for both soils are also drawn showing that both soils swelled and shrunk for almost the whole range of void ratio during the experiment.

The LS soil did not show a consistent shrinkage range. Oppositely the HS soil showed a rather pronounced shrinkage dynamics with void ratio values ranging between 0.99 and 1.6 and a shallow slope across the entire shrinkage range. The resulting slope value <1 implies a lower reduction of the void ratio at a given moisture ratio reduction, thus that shrinkage process is always accompanied by air entry in pores.

Two-Dimensional Image Analysis

In figure 3 porosity values and the pore size distributions (PoSD) of treated and control samples of the two soils are reported. In the graphs the amount of each pore size class is expressed as percentage of the total surface of the analyzed sample. Since images were acquired at pixel resolution of 10 μm, the porosity values refer to all pores larger than 10 μm. The identified PoSD classes have an interval of 20 μm because of the iterative application of the "opening" algorithm with circular "structuring elements" having diameter which increases two pixels per step.

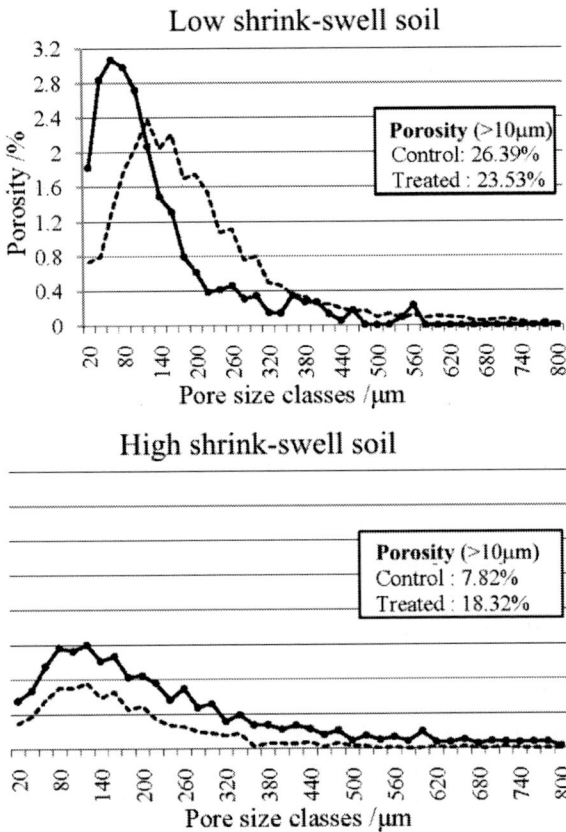

Figure 3. Pore size distributions results. PoSDs of control and samples of low shrink-swell soil and high shrink-swell soil treated with calcium carbonate. Dashed lines refer to control. Porosity values refer to pores larger than image resolution (pixel size).

Porosity values were higher in LS than in HS soil (Figure 3). PoSDs showed the presence of pores having width till 800 μm.

The application of $CaCO_3$ to the LS soil produced a slight decrease in porosity value (26.4% in the control and about 23.5% in the treated samples). PoSD showed that the overall reduction in porosity was a result of the combined effect of a decrease in the larger pore size classes (>120 μm) and an increase in the smaller pore size classes (<120 μm). Calcium carbonate in HS soil caused an increase of porosity value from 7.8% to 18.3% across all pore size classes (Figure 3).

Micromorphological Analysis

Calcium carbonate has been detected in treated soil samples as micrite coatings and micrite segregations micromorphological features that were localized in the pore system. Moreover, $CaCO_3$ finely dispersed in the soil matrix (hereafter named as matrix impregnation) was also considered.

The results of the point counting of these micromorphological features are reported as the percent of the solid phase area of the analyzed thin sections (Table 2). Standard errors were always <0.009, and were therefore not included in the table.

Table 2. Frequencies (%) of micromorphological features expressed in percent area of the solid phase

Soils	Micrite coatings (%)	Micrite segregations (%)	$CaCO_3$ dispersed in soil matrix
Low shrink-swell	1.7	16.6	
High shrink-swell	3.6	1.5	10.7

The application of $CaCO_3$ to LS soil induced the formation of micrite coatings (1.7%) and a high percentage of segregations (16.6%) which formed "bridges" between sand grains in many sites (Figure 4a). In HS soil samples higher percentage (3.6%) of micrite coatings (Figure 4b) and lower percentage (1.5%) of micrite segregations were found compared to LS soil. In the HS soil $CaCO_3$ impregnating the soil matrix were about 10.7%; this micromorphological feature was not detectable in the sandy LS soil.

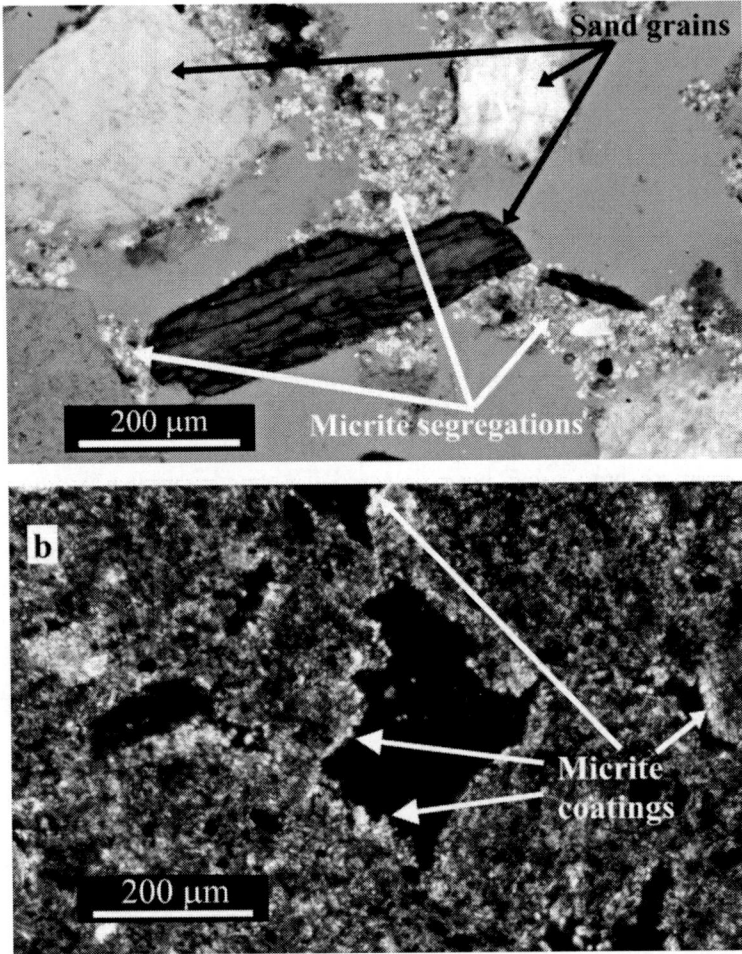

Figure 4. Micrographs of samples treated with calcium carbonate. Micrite segregations in pores in low shrink-swell soil (a) and micrite coatings on pore walls in high shrink-swell soil (b). Photos were acquired (a)with partially cross polarized light (XPL) and (b) with cross polarized light. Pores appear (a) gray and (b) black.

DIFFERENCES BETWEEN SOILS WITH HIGH AND LOW SHRINKING-SWELLING CAPACITY

Soil image analysis allowed overall to quantify the changes of pore size distribution (fig.3) induced by the redistribution of $CaCO_3$ in the tested soils, while soil micromorphological analysis helped to understand mechanisms of such redistribution and provided to identify and quantify the pore-related pedofeatures induced by the addition of this inorganic agent.

To better highlight the differences between soils, comparisons of results between control and treated samples were summarized in the synoptic Table 3.

Table 3. Summary of the quantitative comparison between control and treated samples. Percent variation in porosity values are rounded to the nearest 5%. Changes in Pore Size Distributions (PoSD) are summarized indicating the pore size ranges where porosity has increased or decreased. The most frequently encountered (>1% of solid phase) micromorphological features induced by the treatments are reported with their frequencies

Analysis	Low shrink-swell soil	High shrink-swell soil
Porosity (>10μm)	-10%	+ 135%
PoSD range/ μm	+ (<120), - (>120)	everywhere
Micromorph.	Coatings (1.7%), Segregations (16.6%)	Coatings (3.6%), Segregations (1.5%)

Since it is well known that shrinking-swelling processes increase soil porosity (Scott, 2000; Pires et al., 2008; Horn et al., 1994), the lack of such dynamics in the LS soil can be considered the underlying factor to the reduction in porosity observed in the treated LS samples (Table 3).

The mobilization of $CaCO_3$ in suspension during W/D cycles induced the formation of a high percentage of micrite segregations (Table 2), most likely forming as the coalescence of multiple micrite coatings. These micrite segregations are frequently arranged as bridges between sand grains, thus fragmenting the pore space (Figure 6a). This fragmentation explains the observed decrease of pores >120 μm, and the corresponding increase in <120 μm pores (Figure 3). Overall, there was a 10% reduction of porosity partially due also to the formation of micrite coatings on the sand grain surfaces.

Unlike LS soil, in HS soil $CaCO_3$ induced an evident increase of porosity. The substantial shrink-swell dynamics (Figure 1) of HS soil and its high shrink-swell potential may have been the primary factors driving this increase in about all pore size range.

Micromorphological analysis revealed a high percentage of micrite coatings on the pore walls (Figure 4b). Micrite coatings produced a very porous microstructure as is typically observed in the formation of calcic horizons (Gile et al., 1966; Lal, 2006). In this soil $CaCO_3$ acted as cement and likely stabilized the walls of new pores formed at each successive W/D cycle. This produced a cumulative effect in the pore development with the succession of the cycles, which explains the high level of porosity observed.

CONCLUSION

The results of this experiment on two soils with different shrinking-swelling capacity allowed, by coupling soil micromorphology and image analysis, to (i) quantify the effects of the mobilization of $CaCO_3$ in the soil pore system, (ii) identify the mechanisms by which the resulting pedofeatures affect soil pore development and (iii) measure the consequent changes in pore size distribution.

Micromorphological observations confirmed that $CaCO_3$ added in the soils has mainly moved in suspension during the W/D cycles. Its redistribution in the pore system is clearly linked to the water movement in the soil samples. The fine particle size class of the HS soil in combination with its complex PoSD may determine, in comparison with the LS soil, a larger surface area, a higher water retention capacity and a lower water movement. All these properties can explain the occurrence of calcium carbonate impregnating only the soil matrix of the HS soil.

The lower percentage of micrite coatings observed in the LS soil than HS soil could be explained by its smaller surface area. Moreover, the observed bridges of micrite segregation between sand grains in the LS soil can be explained as the result of the accumulation in the areas where water formed the menisci during the desiccation phase of W/D cycles.

Although the experiment was performed on experimental samples, results can help to understand the contribution of CaCO3 to the mechanisms of soil structure formation in the field. For instance, the increase of all pore size classes induced by micrite coatings that cemented the new pore walls in HS soil and the soil pore space fragmentation produced by micrite segregations in LS soil provide evidences on the importance of mechanical migration of calcium carbonate in the formation of soil structure in the calcic horizons.

In conclusion the used approach allowed to emphasize the large contribution that can be obtained combining quantitative (image analysis) and qualitative soil micromorphology in both process understanding and quantification of some soil properties (e.g. porosity).

ACKNOWLEDGMENTS

We thank Dr A. Basile, Dr R. De Mascellis and Mr B. Di Matteo from CNR-ISAFOM for the help in the study of shrinkage characteristic, the granulometry measures of the soils and the preparation of the soil thin sections, respectively; Dr S. Vingiani and Dr L. Minieri from University Federico II of Naples for the support in mineralogical and chemical analysis of the soils, respectively.

REFERENCES

Baghernejad, M., Dalrymple JB., 1993. Colloidal suspensions of calcium-carbonate in soils and their likely significance in the formation of calcic horizons. *Geoderma.* 58, 17-41.

Brasher, B.R., Franzmeier, D.P., Valassis, V., Davidson, S.E., 1966. Use of Saran resin to coat natural soil clods for bulk density and water retention measurements. *Soil Sci.* 101, 108.

Bronick, C.J. and Lal, R. 2005. Soil structure and management: a review. *Geoderma.* 124, 3 –22.

Cattle, S.R., Farrell, R.A., McBratney, A.B., Moran, C.J., Roesner, E.A., Koppi, A.J., 2000. Solicon – PC Version 1.0. The University of Sidney and Cotton Research and Development Corporation.

Delmas, A.B., Benier, J. & Chamayou, H. 1987. Les figures de corrosion de la calcite: Typologie et sCquence tvolutives. In: *Micromorphologie des Sols* (eds N. Fedoroff, L. M. Bresson & M. A. Courty), pp. 303-308. Association Franqaise pour I'Etude du Sol, Plaisir

Durand, N., Monger H. C., Canti M. G., 2010. Calcium Carbonate Features. In: *Interpretation of Micromorphological Features of Soils and Regoliths* (Eds Georges Stoops;Vera Marcelino, Florias Mees). Elsevier, Amsterdam, The Netherlands, pp. 149-154.

Falsone, G., Catoni, M., Bonifacio, E. 2010. Effects of calcite on the soil porous structure: natural and experimental conditions. *Agrochimica.* 54, 1 -12.

Fitzpatrick, E.A. 1993. *Soil Microscopy and Micromorphology.* Wiley, Chichester.

Gargiulo, L. 2008. Indagini innovative sulla strutturazione del suolo. *Graduation thesis in Applied Pedology*, University Federico II of Naples.

Gee, G.W. and Or, D., 2002. Particle-size analysis. In: Dane, J.H., Topp, G.C. (Eds.), *Soil Science Society of America Book Series*: vol. 5, Methods of Soil Analysis. Part 4. Physical Methods. Madison, WI, pp. 255-293.

Gile, L.H., Peterson, F.F., Grossman, R.B. 1966. Morphological and genetic sequences of carbonate accumulation in desert soils. *Soil Sci.* 101, 347-360.

Groenevelt, P.H. and Grant, C.D. 2004. Analysis of soil shrinkage data. *Soil Till. Res.* 79, 71-77.

Hillel, D. 1980. Fundamentals of soil physics. Academic Press, New York.

Horgan, G.W. 1998. Mathematical morphology for analyzing soil structure from images. *Eur. J. Soil Sci.* 49, 161-173.

Horn, R., Taubner, H., Wuttke, M., Baumgartl, T. 1994. Soil physical properties related to soil structure. *Soil Till. Res.* 30, 187-216.

Jaillard, B., Guyon, A. & Maurin, A.F. 1991. Structure and composition of calcified roots, and their identification in calcareous soils. *Geoderma,* 50, 197-210.

Kaemmerer, M. & Revel, J.-C. 1996. New data on the laminar horizon genesis of calcretes developed on Morocco coarse Quaternary alluvium: Consequences on the desertification process. *Arid Soil Research and Rehabilitation,* 10, 107-123.

Kaemmerer, M., Revel, J.-C. & Barlier, J.-F. 1991. Formation des amas friables et des nodules calcaires dans des sols argileux en rkgions temptrte et semi-aride. *Science du Sol,* 29, 1-12.

Khormali, F., Abtahi, A., Stoops, G., 2006. Micromorphology of calcitic features in highly calcareous soils of Fars Province, Southern Iran. *Geoderma.* 132, 31-46.

Lal R., 1991. Soil structure and sustainability. *J. Sustain. Agric.* 1, 67-92.

Lal, R. 2006. *Encyclopedia of Soil Science*. Second Edition, Vol I, Vol II. Taylor and Francis, Boca Raton, FL.

Loeppert, R.H., Suarez, D.L. 1996. Carbonate and gypsum. In: Sparks, D.L. (ed.), *Methods of Soil Analysis*, Part 3, Chemical Methods SSSA Book Ser. 5. American Society of Agronomy and Soil Science Society of America, Madison, Wisconsin, USA.

Loveday, J. 1972. Field aspects of swelling and shrinking. In: *Physical aspect of swelling clay soils*. A Symposium at the Univeristy of New England, February 1972. Department of University Extension, University of New England, Armidale, NSW.

McKeague, J.A., Guertin, R.K., Valentine, K.W.G., Belisle, J., Bourbeau, G.A., Howell, A., Michalyna, W., Hopkins, L., Page, F.. Bresson, L.M.

1980. Estimating illuvial clay in soils by micromorphology. *Soil Sci.* 129, 386-388.

Mele, G., Basile, A., Leone, A.P., Moreau, E., Terribile, F., Velde, B. 1999. The study of soil structure by coupling serial sections and 3D image analysis. Modelling of transport processes in soils. In: Workshop of EurAgEng's Field of Interest on Soil and Water (ed. Feyen and K.Wiyo), pp. 103-117. Leuven.

Muneer, M. and Oades, J.M. 1989. The role of Ca-organic interactions in soil aggregate stability.1. Laboratory studies with glucose-C-14, CaCO3 and CaSO4.2H2O. *Aust. J. Soil Res.* 27, 389-399.

Murphy, C.P. 1983. Point counting pores and illuvial clay in thin section. *Geoderma.* 31, 133-150.

Pagliai, M. Vignozzi, N. 2002. Image analysis and microscopic techniques to characterize soil pore system, in: Blahovec, J., Kutílek, M (Eds.), *Physical methods in agriculture: approach to precision and quality.* Kluwer Academic Plenum Publishers, New York, pp. 13-38.

Peech, M. 1965. Hydrogen-ion activity. In: Black, C.A. (Ed.), *Methods of Soil Analysis, Part 2.* American Society of Agronomy and Soil Science Society of America, Inc. Madison, Wisconsin, USA.

Pires, L.F., Cooper, M., Cássaro, F.A.M., Reichardt, K. Bacchi, O.O.S., Dias, N.M.P. 2008. Micromorphological analysis to characterize structure modifications of soil samples submitted to wetting and drying cycles. *Catena.* 72, 297-304.

Rhoades, J.D. 1996. Salinity: Electrical conductivity and Total Dissolved Solids. In: Sparks D.L. (Ed.), *Methods of Soil Analysis, Part 3*, Chemical Methods SSSA Book Ser. 5. American Society of Agronomy and Soil Science Society of America, Madison, Wisconsin, USA.

Scott, H.D. 2000. *Soil physics: agricultural and environmental applications.* Iowa State University Press, Iowa.

Serra, J. 1982. *Image Analysis and Mathematical Morphology.* Academic Press, London.

Shankar, N., Achyuthan, H., 2007. Genesis of calcic and petrocalcic horizons from Coimbatore, Tamil Nadu: Micromorphology and geochemical studies. *Quatern. Inter.* 175, 140-154.

Six, J., Bossuyt, H., Degryze, S., Denef, K. 2004. A history of research on the link between (micro)aggregates, soil biota, and soil organic matter dynamics. *Soil Till. Res.* 79, 7-31.

Soil Survey Division Staff. 1993. Soil survey manual. *Soil Conservation Service Handbook 18, chapter 3.* U.S. Department of Agriculture.

Summer, M.E. and Miller, W.P. 1996. Cation exchange capacity and exchange coefficients. In: Sparks, D.L. (Ed.), *Methods of Soil Analysis, Part 3, Chemical Methods SSSA Book Ser. 5.* American Society of Agronomy and Soil Science Society of America, Madison, Wisconsin, USA.

TRANSVALOR 2000. Micromorph Version 1.4 CMM. Ecole des Mines, Armines.

U.S.D.A., 2004. Soil Survey Laboratory Methods Manual. *Soil Survey Investigations Report.* No. 42,Version 4.0.

U.S.D.A., 2010. Department of Agriculture, Natural Resources Conservation Service. National Soil Survey Handbook, title 430-VI. Available online at http://soils.usda.gov/technical/handbook/.

U.S.D.A.,1954. Determination of the Properties of Saline and Alkali Soils in US Salinity Laboratory Staff, Diagnosis and Improvement of Saline and Alkali Soils. *Agriculture Handbook* n. 60. L.A. Richards, Editor. pp. 25-30.

Velde, B. 1999. Structure of surface cracks in soil and muds. *Geoderma.* 93, 101-124.

Walkley, A. and Black, I.A. 1934. An examination of the Degtjareff method for determining organic carbon in soils: Effect of variations in digestion conditions and of inorganic soil constituents. *Soil Sci.* 63, 251-263.

Wilson, M.J. 1987. *A Handbook of Determinative Methods in Clay Mineralogy.* Chapman and Hall, New York.

In: Carbonates
Editors: B.A. Hughes, T.C. Wagner

ISBN: 978-1-62948-178-4
© 2013 Nova Science Publishers, Inc.

Chapter 4

TYPES OF PETROLEUM RESERVOIRS IN CARBONATE SEDIMENTS OF THE RUSSIA BASINS

Vitaly G. Kuznetsov[*]
Gubkin's Russian State University of Oil and Gas, Russia

ABSTRACT

In petroleum basins of the Russian Federation oil and gas fields in carbonate reservoirs have been discovered in rocks ranging from the Riphean to the Eocene. giant fields, are controlled by reefs. Depending on the paleoclimatic zone, the seals are composed of salt or, rarely, of shale. The largest amount of the fields are found in cratonic carbonate formations deposited under arid climatic conditions. Regional seals are formed by salt, anhydrite, and dolomicrite. Multilayer reservoirs predominate, but massive reservoirs are also common. The distribution of reservoir types and their quality are strongly uneven. A large number of fields, including giant fields, are controlled by reefs. Massive reservoirs predominate, but the distribution of porosity and localization of zones of improved reservoir properties are variable and controlled by the morphogenetic types of the reefs- Carbonate formations deposited under humid climatic conditions contain much less hydrocarbon reserves. The seals are generally composed of shale. The reservoirs are stratal, rarely multilayer. The fields are commonly small. A number of fields, some of

[*] E-mail: vgkuz@uyandex.ru.

them highly productive, are present in Upper Cretaceous carbonate rocks of the North Caucasus region. The carbonates consist of remain of planktonic organisms. Seals for the hydrocarbon pools are composed of shale. The reservoirs are massive and layered-massive. Fractures and stylolites play a leading role in controlling the reservoir properties.

Keywords: Carbonate sediments, oil and gas, reservoir, Russian Federation

INTRODUCTION

General Information on Oil and Gas Presence in Carbonate Deposits of Russian Federation

Carbonate deposits are widely located within the oil and gas bearing basins on the territory of Russian Federation. However, the level of oil and gas reserves connected with carbonate reservoirs is considerably lower than the word wide values.

They contained about 8% of resources in the 1980s, but to end of century this value rose to 12,8% (Belonin et al., 2005). Only 11 of 118 fields with recoverable reserves more than 30 billion tons contain petroleum in carbonate sediments.

Stratigraphic productively range of carbonate reservoirs is rather wide - from Middle-Upper Proterozoic (Riphean) deposits of the East-Siberian craton - ancient platforms (Yurubcheno-Takhomskaya zone) to Eocene (North PriCaucasus).

Main reserves time are connected to the carbonate reservoirs of Paleozoic, where 94 – 95% of general reserves of carbonate sediments are concentrated. Paleozoic carbonate contain a little more than 50% reserves of this Erathem. About 2 – 3% reserves are related to in the Proterozoic and Mesozoic, and least of all – less 1% to Cenozoic reserves of each Eraterms.

This situation explains rather small to compare with the world date oil-gas content of carbonate sediments in Russian Federation. Maximum of world reserves of hydrocarbons fall on Mesozoic, when territories of contemporary Russia were situated in the zone of cold climate, unfavourable for carbonate sedimentation. This define primary petroliferous character of terrigenous rocks of the West Siberia basin. A small rise of portion of carbonate in last years is stipulated by discovery of new fields in Proterozoic and Cambrian of East Siberia and relative reduction in terrigenous deposits of the West Siberia.

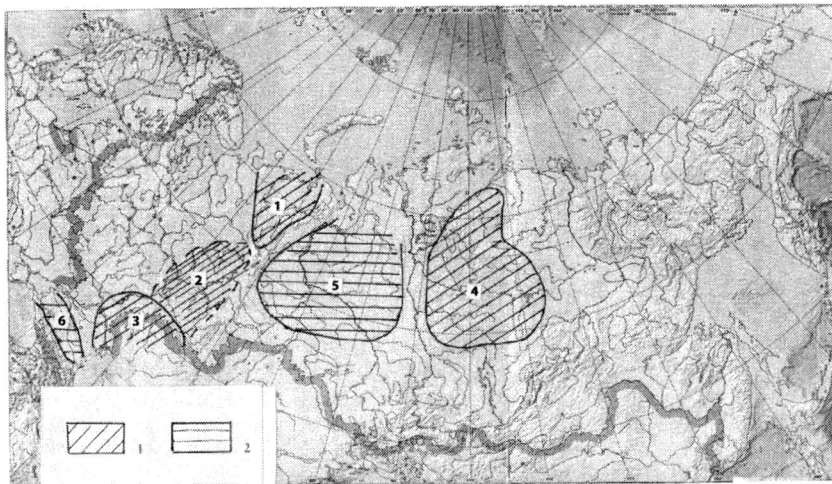

Figure 1. Oil and gas province of the Russian Federation contained hydrocarbons in the carbonate deposits.

Productively distribution by regions, and tectonic elements is rather nonuniform. The main basins where carbonate reservoirs determine the main oil and gas presence in the sedimentary cover are Timan-Pechora, Volga-Urals, PriCaspian, and Lena-Tunguska and North Caucasus a lesser extent (figure 1).

Brief Summary of Oil and Gas Presence in Carbonate Deposits of Separate Provinces and Regions

1.   Ancient platforms - cratons

Oil and gas presence in carbonate reservoirs of the East-European platform is the main object which determines at present oil and gas reserves in carbonate deposits of the Russia.

To one degree or another the carbonate deposits are productive from the Middle Ordovician (Kosju-Rogovsk depression of the Timan-Pechora basin) to Kazanian stage of the Upper Permian (South-East Volga-Urals basin), however, the quantitative distribution of reserves is rather nonuniform. Almost a quarter of all the hydrocarbon reserves of the platform are located in the Upper-Viseian-Bashkirian carbonate complex. Approximately the same amount of reserves are located in Middle-Carboniferous - Lower Permian carbonate complex. Upper-Devonian-Tournaisian carbonate deposits contain of reserves. The resources of other deposits are rather modest (Gabrielyantz et al., 1989).

The main productivity is connected with zones of active downwarping of eastern part of the platform - large and deep Pechora and PriCaspian depressions (syneclises) including Timan-Pechora and PriCaspian basin and the Volga-Urals anteclise where the like oil and gas bearing basin is located.

In the Timan-Pechora basin the carbonate reservoirs contain about 64% of hydrocarbone reserves of this basin (Belonin et al., 2005). Here the commercial oil and gas presence of more ancient within the East-European platform of the Middle-Ordovician- Lower Devonian carbonate deposits has been determined. Besides the Upper-Devonian-Tournaisian and Visean-Lower Permian carbonate complexes are productive. In the Volga-Urals basin the commercial oil fields have been found in the carbonate beds of the Middle Devonian siliciclastic-carbonate complex. The most important are thick carbonate complexes – Upper Devonian-Tournaisian, Visean-Bashkirian, Middle-Upper Carboniferouse- Lower Permian. The youngest ones are small in reserves pools in carbonate deposits of the Kalinovsk Siute of the Kazanian Stage of the Upper Permian located in the southern and south-eastern parts of the province. About 44% of oil and over 80% of gas of total reserves are found in the carbonate rocks, of the basin (Gabrielyantz et al.,1989 , Belonin et al., 2005). The number of fields in the carbonate deposits of the PriCaspian basin is considerably small but many of them are very large and it mainly determines their greatest economic value, but they locate in Kasachstan. The carbonate rocks of Upper Proterozoic - Vendian and Cambrian are widely developed within the Siberian platform and their role as oil and gas reservoirs is very essential. The carbonate deposits of the Lena-Tunguska basin in the south of the platform are productive. The dolomites of the Middle-Upper Proterozoic (Riphean) of the Yurubcheno-Takhomsk zone and salt-carbonate formations of the Vendian-Cambrian are productive.

2.   Young (Epihercynian) platforms.

The productivity of carbonate deposits has been determined within two basins of young platforms - West-Siberian and North Caucasus. Oil and gas presence in carbonate formations of the West Siberian basine known for its unique resources is extremely small. Paleoclimatic conditions caused in Mesozoic and Cenozoic of this basin only accumulation of clastic rocks which are the main oil and gas reservoirs. Only some places of Paleozoic (Devonian) deposits of the preplatform - basement complex have a few of small hydrocarbon pools. Rather wider is a stratigraphic range of oil and gas presence in carbonate formations of the North Caucasus basins confined to the Skyphian platform. Commercial oil fields have been found in the Lower and Middle Triassic of Eastern PriCaucasus. Separate pools are known in the carbonate formations of the Upper Jurassic and Paleogene. The main formation, where the carbonate rocks are productive, is the Upper Cretaceous carbonate complex of the Eastern PriCaucasus and Tersko-Caspian trough.

Types of Oil and Gas Bearing Carbonate Formations

Even a short summary of carbonate reservoirs shows a wide stratigraphic range of their development, relation with different tectonic structures and, mainly their connection with different formation types (Table 1).

Table 1. Oil and gas basins of the Russian Federation, contained hydrocarbons in the carbonate sediments

Tectonic position	Oil and gas basins	Age	Type of carbonate sediments
Ancient platform-craton	Timan-Pechora	$C_1v - P_1$ $D_3 - C_1t$ $O_2 - D_1$	Platform semiarid Platform humid, Reef Platform humid
	Volga-Urals	P_2kaz $C_2 - P_1$ $D_3 - C_1t$ $C_1v - C_2b$ $D_3 - C_1t$ D_2	Platform arid Platform arid, Reef, Bituminous-silicious-clay-carbonate , Platform arid Platform humid, Reef, Bituminous-silicious-clay-carbonate Reef, Siliciclastic-carbonate
	PriCaspian	$C_2 - P_1$ $C_1v - C_2b$	Platform arid, Reef Reef
	Lena-Tunguska	$V - \in$ PR_{2-3}	Platform arid, Reef Platform humid (?)
Young Epihercynian platforms	West Siberia	D	Isolated platform
	North Caucasus	P_{1-2} K_2 J_3 T_3 T_1	Siliciclastic-carbonate Planctonogenic Platform arid Platform humid (?) Reef

As it is accepted in Russian literature on geology, the formation in this case a geological body representing both natural and regular combination of rocks integrtated by general conditions of formation and appearing at definite stages of development of main structural zone of the Earth crust (Khain, 1973). Thus, the formations are separated both by petrographic type of the making up rocks and accumulation conditions (tectonic, paleogeographic, paleoclimatic). Purely carbonate deposits are more often and a maximum degree productive while to a lesser degree salt-carbonate and to a considerable extent the carbonate-siliciclastic are productive.

From the point of the origin the oil and gas presence has been determined in various formations. The most abundant are the platform bentogene carbonate and salt-carbonate formations-In this case the term "platform" is used as tectonic and for PreCambrian structures partial correspond to "craton".

In English literature it is in accordance with two types of carbonate platforms as morphological and geomorphological notions - epeiric platform and more seldom, rimmed shelf (Tucker and Wright, 1990).

Among similar formations of the arid climatic zone the productive are the Vendian-Cambrian salt-carbonate deposits of the Lena-Tunguska basin of the Siberian platform, the Visean-Bashkirian and Middle Carboneferous-Lower Permian of the Timan-Pechora and Volga-Urals basin, the Kazansk of the north of the Volga-Urals province.

The platform carbonate formations of a humid zone are considerably more seldom. The studied examples are the Upper Devonian – Tournaisian formations of the Timan-Pechora and Volga-Urals basin. It's not improbable that they are close to the Middle-Ordovician – Lower Devonian deposits of the Timan-Pechora basin and the Middle Triasic of the Eastern Pricaucasus.

Oil and gas presence in the formation of the isolated carbonate platforms and reefs is rather widespread and important. They are related with the Lower Cambrian reefs of the Eastern Siberia, the Upper Devonian of the Timan-Pechora and Volga-Urals basin, biostrome buildups of the Cambrian Lena-Tunguska and Carboniferous of Volga-Urals, the Carboniferous - Lower Permian reefs of the Volga-Urals and PriCaspian basin, the Lower Triassic reefs of the Eastern Pricaucasus. In Pricaspian only known example of the productive isolated carbonate platform is likely the Middle Carboniferous of the Astrakhan gas-condensate field. The fourth type of purely carbonate formations is the planctonic formation of the Upper Cretaceous, being productive within the East Pricaucasus. The carbonate members of the carbonate-siliciclastic formations are productive in the Middle Devonian of

the Volga-Ural basin, the Eocene of the Pricaucasus. The bitumenous-
siliceous - argilaceous- carbonate rather deep-sea formation of
noncompensated depression (starved basin) is specific. The commercial oil
and gas bearing deposits are the Upper Devonian and in some places the
Upper-Devonian-Tournaisian deposits of a number of regions of the Timan-
Pechora and Volga-Urals basin where in some areas the commercial inflows of
oil and gascondensate as well as the Lower-Permian deposits of the Cis Ural
foredeep of the Volga-Urals basin have been achieved.

The Types of Carbonate Reservoirs
and Their Interrelation with the Formations

The carbonate deposits of oil and gas bearing provinces of the Russian
Federation three main types of reservoirs, namely, massive stratal and
lithologically isolated have been determined In this case in Russian literature
the term "natural reservoir" is used in a wider sense. Natural reservoir is a
geologic body which consists of a number of layers (in a particular case of one
layer) possessing reservoir properties (porosity and permeability); this body is
sealed by practically impermeable rocks, it can contain circulating fluids. The
most important property of a reservoir, i.e. fluids distribution peculiarities,
character of their in-reservoir migration, is defined by the two parameters:
spatial interrelation of rocks with different reservoir properties, i.e. inner
structure, composition of a geological body, as well as by porosity and
permeability of the elements it consists of (Kusnetsov, 1989, 1992). Natural
reservoir may be of different sizes and a general case is distributed along the
territory of the whole basin. In this case oil and gas pools occupy only part of
the reservoir in the location where there is a possibility of trapping and
conservation of hydrocarbons presented by on anticlinal fold, erosional outlier,
lithological replacement or stratigraphic nonconformity. Massive reservoirs
imagine of thickness sufficiently homogeneity rock mass (massive reservoir is
relatively uniform thickness powerful) and the fluid migration equally takes
place both along lateral and vertical section. It's quite natural that the reservoir
thickness can't be compared with its areal extent, so the uniformity of
migration is displayed in equal ability to migration along lateral and vertical
but its distance (figure 2). In strata reservoirs and their varieties – multilayered
the lateral migration prevails and the vertical one is restricted by the thickness
of the permeable layer. In multilayer reservoirs the layers with commercial
values are separated by the low-permeable ones. In the latter the filtration is

rather difficult and takes place when pools are formed and it causes a single water-oil and gas-water contact. Commercial production is obtained only from permeable layers. Lithologically restricted or locally developed isolated reservoirs are characterized by an abnormal shape, relatively limited sizes. Interreservoir migration partially takes place due to rocks pressure, partially due to the energy of hydrocarbons themselves. Solution gas drive is typical, in particular. The reservoirs of one type may occur in various formations, but one or another type of reservoirs is clearly observed in some definite formations. In arid platform formations clearly prevail stratal and particularly multilayer reservoirs. Salts, seldom anhydrites, argillaceous dolomicrits with anhydrites often serve as seals. In this case a clear asymmetry of formation structure and the reservoir associated with it is marked (figure 3). The best reservoir properties are observed in that facial zone of formation that is characterized by the average ocean salinity. With a considerable reservoir thickness and section uniformity the reservoir record in this zone may acquire the character of a massive one. In a distal part of the formation in the zone of increased salinity the reservoir properties are low and the multilayer reservoir are formed. The scheme of the distribution of hydrocarbon pools in these deposits are shown as example (Figure 4). Pay attention to great stratigraphic range of the reservoir and the pool of the Orenburg field and the availabilityof several pools on other fields located north-west.

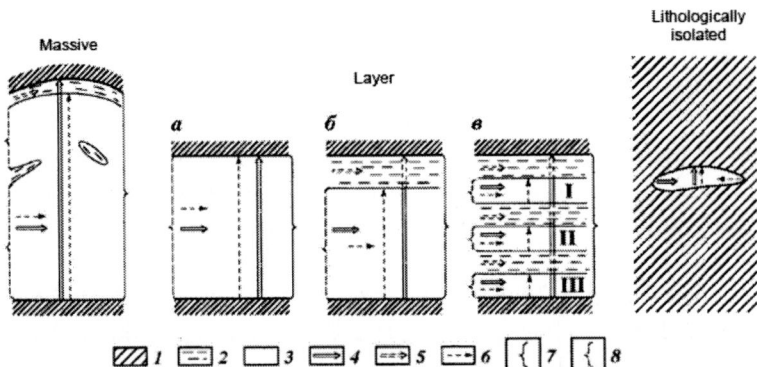

Figure 2. Principal charts of structures of natural reservoirs. 1 – Seals; 2 – Semi-seals; 3 – Permeable rocks; Possible filtration of fluids: During pool formation: 4 – Intensive; 5 – Impered; 6 – Under development; 7 – Volume of reservoir; 8 – Net thicknesses. I, II, III – Productive beds.

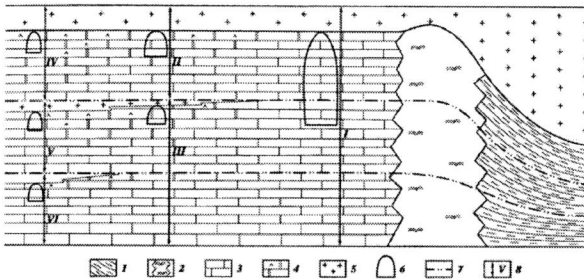

Figure 3. Scheme showing the relations of facies and stratigraphic volume of reservoirs in platform carbonate formation of the arid climatic zone. 1 – Sediments of naturally-salted pools connected with the World Ocean; 2 – The zone of possible reef's development; 3 – The priority development area of limestones including organic ones; 4 - The priority development area of micritic limestones and dolomites often sulphatized; 5 – Sulphate and halogen sediments; 6 – Oil and gas pools; 7 – The chronostratigraphic levels; 8 – The capacity of oil and gas reservoirs.

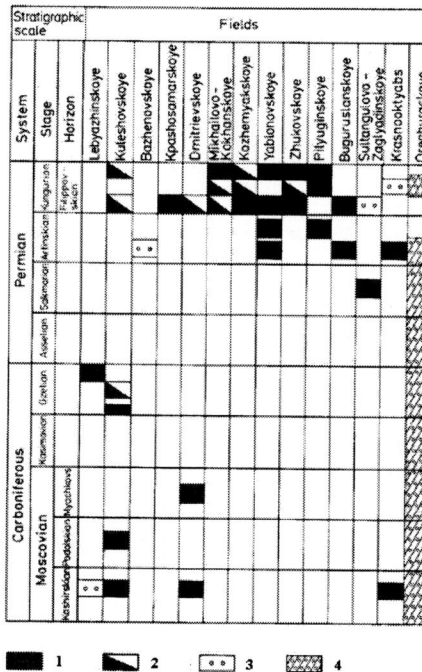

Figure 4. Distribution of hydrocarbon accumulations in the Middle Carboniferous – Lower Permian deposits in some areas of the East-European platform. 1 – Oil; 2 – Oil and gas; 3 – Gas; 4 – Gas-condensate.

Multilayer reservoirs take place in the Cambrian salt-carbonate deposits of Lena-Tunguska basin (Kuznetsov, 1995). Multilayer containers are available in the Lena-Tunguska basin.

In humid platform formations single layer reservoirs with one or other lensl-ike heterogeneity are developed. In this case the space changes of the structure and reservoir properties are not very great. Clays are considered seal screening members.

Structure of section of reservoirs is different for different climatic types of formations and this are stipulated of various structure of cycles.

Shallow-water carbonate section are always characterized by cyclic structure (Kuznetsov, 2006). Moreover, composition and structure of cycles are different in humid and arid climatic zones. The first most remarkable and, at first sight, readily explainable difference consists in the composition of cycles.

They are dominated by limestones in humid zones and dolomites in arid zones. The second difference lies in the structure of cycles. In carbonate sequences of arid zones, cycles have the three-member structure: their basal member is composed fine-grained, frequently clayey dolomites and dolomitic marls; the central (middle) member consists of granular organogenic-detrital limestones with irregular dolomitization in some places; and the upper member is again composed of microgranular microbial dolomites, (including stromatolitic varieties).

Cycles of humid zones have a simpler (two-member) structure: microgranular limestones (with different contents of clays) in the lower part of the section and granular limestones usually with diverse organic remains in the upper part. In both cases, boundaries of cycles are relatively sharp, sometimes with erosion signs.

The upper surface of cycles in humid zones frequently bears traces of karstification and leaching. These features of cycles are also responsible for different physical properties (porosity and others) of rocks. In arid zones, the highest porosity is typical of central parts of cycles.

In cycles of humid zones, such properties are characteristic of their middle and, particularly, upper parts, where primary pores are accompanied by leaching caverns.

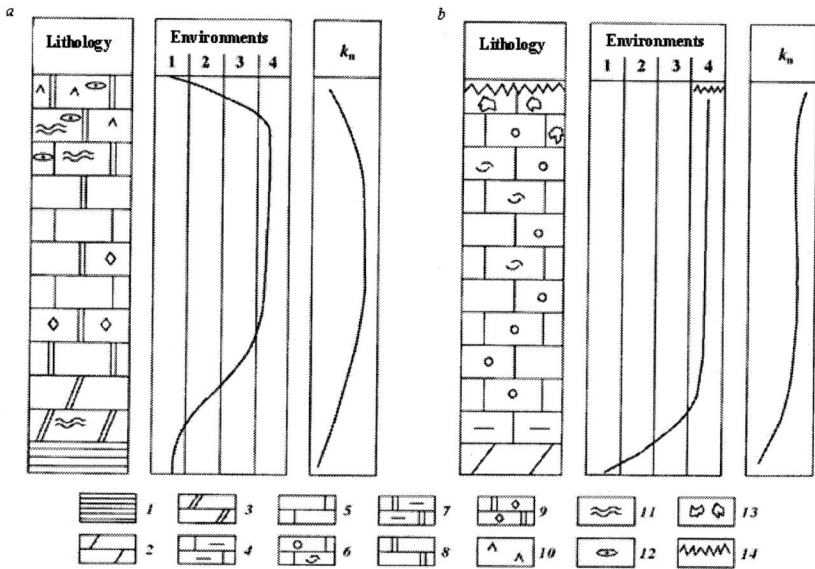

Figure 5. Principal scheme of structure cycles of shallow-water carbonate deposits of arid (a) and humid (b) climatic zones. Environmtnts: 1 – supralittoral; 2 – littoral; 3 – sublittoral; 4 – shallow-water. Rocks: 1 – shale; 2 – marl; 3 - dolomitic marl; 4 – clayey dolomite; 5 – micritic limestone; 6 – greenstone; 7 – clayey dolomite; 8 – micritic dolomite; 9 – crystal dolomite; 10 m- anhydrite. Texture: 11 – microbial (laminite, stromatolite); 12 – fenestral; 13 – cavern of subaeral leaching; 14 – interruption/. Kp – principal scheme change of porosity.

In connection (Therefore) with this in arid formations prevail (predominate) multi-layered reservoirs, and in humid– one-layered one. In planktonogenic formations the primary reservoir properties especially permeability are not great, but an intensive fracturing unites thickness sections with a possible considerable vertical migration into a single hydrodynamic system. It determines mainly the massive type of a reservoir when the oil pools cavers the whole rather a considerable thickness of the carbonate deposit. In known examples of these formations clays serve as screening deposits. Lithologically restricted reservoirs in locations of porosity and permeability development are associated with bituminous-siliceous argillaceous carbonate formations. The formation of the porosity zones is confined to the sections of diagenetic dolomitization and fracturing. Reservoirs of reef formation differ in a variety (Kuznetsov, 1997). At the same time caverns and intercrystal pores formed by dolomitization are available along with an ordinary type of pore space-growth framework porosity. As to common massive type, the details of the structure are different for separate morphologo-genetic reef types (figure

6). With a relatively slight vertical heterogeneity the simpliest type of a massive reservoir is observed in considerable small dome-shaped reefs (the Cambrian of the Lena-Tunguska basin, Middle Devonian of the western part of the Pricaspian depression and others). In the reefs with high thickness and developing for a long time, the reservoirs reveal stratal and lens-stratal distribution of the zones with improved reservoir properties (the Permian reefs of the Cis Ural foredeep and others). Different is as well an areal extent of the zones with improved reservoir properties. In dome-shaped reefs they are located in the central part of the reefs (the Permian reefs of Cis Ural and others). In atoll-like reefs the zone of improved reservoir properties is connected with bioherm reef facies and has a ring-shaped form. In asymmetric reef systems surrounding the deepsea depression boards, the zone of improved reservoirs is moved to the basin reef slope (the Pashor reef of Timan-Pechora basin and others). The reef reservoirs are usually screened by salts, more seldom by antihydrites, clays and siliceous-carbonate -argillaceous deposits (the Middle Carboniferous reefs of the Pricaspian basin.

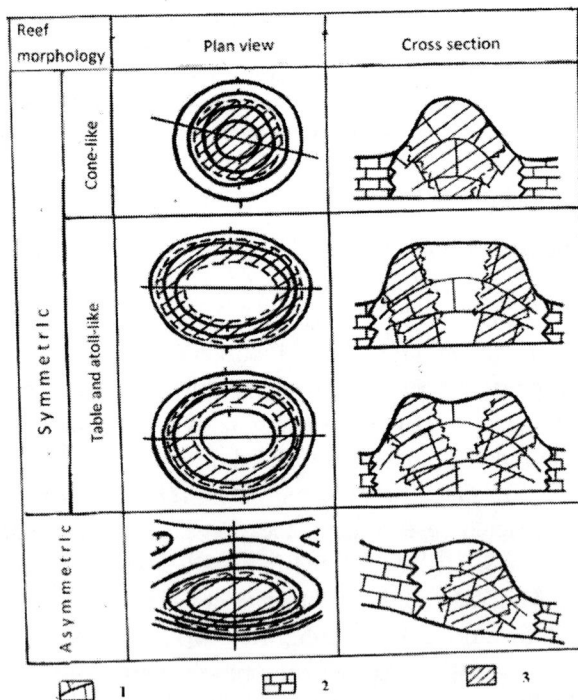

Figure 6. Structures of natural reservoirs of reefs of different morphologic and genetic types. 1 – Reef; 2 – Non-reef sediments; 3 – Area of best reservoir properties.

CONCLUSION

The analysis of oil and gas occurence in the carbonate deposits of the Russia shows that they are productive within various tectonic structures. On ancient platforms they are marginal (Pechora and Pricaspian on the East-European platform) and interplatform (Lena-Tunguska on the Siberian platform) syneclises, antedises (the Volga-Urals in the East-European, Nepsk-Botuobinsk and Baikitsk on the Siberian platform), foredeeps (the Cis Ural on the East European platform). On young platforms these are elongated deep depression (the North-Caucasus). The most importante are the marginal zones of active dawnwarping of ancient platforms adjoining the ancient passive margin of the continents,

The thick carbonate series possess largest reserves, while the carbonate layers confined in the strata of other petrographical rock types have a considerably smaller potential. Among the carbonate formations the leading are the platform ones of the arid zone (mainly epeiric platform) and the reef formations. The latter ones are distinguished by the greatest net thickness (up to 70-80% against 30-40t% in other types of formations) and the maximum reserve density. The most frequent and most important screening deposits are salts. The geological structure of the territory of the Russia paleogeographical and, in particular, paleoclimatic conditions of sedimentation brought about intensive carbonate accumulation in the Paleosoic and the corresponding greatest productivity of the very Paleosoic carbonate deposits.

REFERENCES

Belonin M.D., Belonovskaya L.G., BulachM.Kh., Gmid L.P., Shimansky V.V. 2005. *Carbonate rocks – reservoir rocks oil- gas bearing basins of Russia and adjacent territory.* SPb, Nedra. 260 p. (in Russian).

Gabrielyants G.A., Dikenshteyn G.Kh., Lodshevskaya M.I., Rasmyshlyaev A.A. 1989- Main hydrocarbon fields location in the USSR and foreign countries. Geology, methods of prospecting and exploration of oil and gas fields, N. 8. Moscow, *Allunion Research Institute of Economy of Mineral Products and Exploration* (in Russian).

Khain V-E. 1973. *General geotectonic.* 2nd ed. Moscow, Nedra, 283 pp. (in Russian).

Kuznetsov V.G.1992. Natural reservoirs of oil and gas in carbonate sediments-Moscow, *Nedra*, 240 pp (in Russian).

Kuznetsov V.G. 1995. *Vendian to Cambrian carbonate reservoir of the Siberian Platform //Petroleum Geoscience.* V. 1. Pp. 271 – 278.

Kuznetsov V.G.1997. *Oil and gas in reef reservoirs in the former USSR // Petroleum Geoscience.* V. 3. Pp. 65 – 71.

Kuznetsov V.G. 2006. *Ciclicity of Shallow-Water Carbonate Sediments in Different Climatic Zones // Lithology and Mineral Resources.* V. 41, N 6. Pp. 505 – 517.

Tucker M.F., Wright V.P. 1990. *Carbonate Sedimentology.* Blackwell Oxford. 482 pp.

INDEX

D

T

tantalum, 5
techniques, 8, 71
technologies, 2
TEG, 5
TEM, 18
temperature, 3, 4, 5, 8, 9, 13, 33
territory, 74, 79, 85
tetraethoxysilane, 39
texture, 59, 62
TGA, 8
tin, 23
toxicity, 33
transformation(s), 33, 39, 46
transition metal, 25, 43
transition temperature, 40, 41
transparency, 47
transport, 6, 71
transport processes, 71
treatment, 35, 60, 62
trypsin, 41

U

U.S. Department of Agriculture, 71
uniform, 14
urea, 32, 33, 42, 43
urethane, 39, 40, 41, 42
USA, 70, 71, 72
USDA, 60, 62, 63
USSR, 85, 86
UV, 6, 60

V

vacuum, 60
valence, 8
valve, 11
variations, 72
varieties, 79, 82
viscosity, 40, 60

W

water, viii, 4, 9, 36, 38, 42, 55, 58, 59, 60, 68, 69, 80, 82, 83
wetting, viii, 56, 57, 58, 71
wetting-drying cycle, 56
Wisconsin, 70, 71, 72

X

XPS, 29
X-ray diffraction (XRD), 14, 15, 59
XRD, 8, 10, 12, 14, 29

Y

yield, 4, 5, 6, 7, 22, 35, 36, 37, 42, 45

Z

zirconia, 5, 7, 23, 24